INTRODUCTORY EXPERIMENTS ON BIOMOLECULES AND THEIR INTERACTIONS

INTRODUCTORY EXPERIMENTS ON BIOMOLECULES AND THEIR INTERACTIONS

ROBERT K. DELONG
Kansas State University
Manhattan, KS, USA

QIONGQIONG ZHOU
Missouri State University
Springfield, MO, USA

AMSTERDAM • BOSTON • HEIDELBERG • LONDON
NEW YORK • OXFORD • PARIS • SAN DIEGO
SAN FRANCISCO • SINGAPORE • SYDNEY • TOKYO
Academic Press is an imprint of Elsevier

Academic Press is an imprint of Elsevier
32 Jamestown Road, London NW1 7BY, UK
525 B Street, Suite 1800, San Diego, CA 92101-4495, USA
225 Wyman Street, Waltham, MA 02451, USA
The Boulevard, Langford Lane, Kidlington, Oxford OX5 1GB, UK

Notices
Knowledge and best practice in this field are constantly changing. As new research and
experience broaden our understanding, changes in research methods, professional practices,
or medical treatment may become necessary.

Practitioners and researchers must always rely on their own experience and knowledge
in evaluating and using any information, methods, compounds, or experiments described
herein. In using such information or methods they should be mindful of their own safety
and the safety of others, including parties for whom they have a professional responsibility.

To the fullest extent of the law, neither the Publisher nor the authors, contributors, or editors,
assume any liability for any injury and/or damage to persons or property as a matter of
products liability, negligence or otherwise, or from any use or operation of any methods,
products, instructions, or ideas contained in the material herein.

ISBN: 978-0-12-800969-7

British Library Cataloguing-in-Publication Data
A catalogue record for this book is available from the British Library

Library of Congress Cataloging-in-Publication Data
A catalog record for this book is available from the Library of Congress

For Information on all Academic Press publications
visit our website at http://store.elsevier.com/

Typeset by TNQ Books and Journals
www.tnq.co.in

Printed and bound in the XXX

Working together
to grow libraries in
developing countries

www.elsevier.com • www.bookaid.org

Contents

Introduction   ix

About the Authors   xiii

Master Materials List   xv

Format of the Lab Notebook and Formal Lab Report   xxi

Lab Safety and Policies   xxv

1. An Introduction to Basic Math and Operations in the Biomolecular Laboratory

Safety and Hazards   1
Introduction   1
Technical Review   2
Equipment Highlight   6
Experiment 1   8
Methods and Procedure   8
Results and Discussion   9
Problem Set   10
Additional Materials   10

2. Preparing Buffers at a Specific Molarity and pH

Safety and Hazards   13
Introduction   13
Technical Review   14
Equipment Highlight   16
Experiment 2   17
Results and Discussion   18
Problem Set   18
Additional Materials   18

3. Investigating the Physico-Chemical Properties of Amino Acids and Their Analysis by Thin Layer Chromatography

Safety and Hazards   21
Introduction   21
Technical Review   25
Equipment Highlight   26
Experiment 3   27
Methods and Procedure   28
Results and Observations   31

Discussion   31
Problem Set   32
Acknowledgments   33
Additional Materials   33

4. Rapid Purification, Gel Electrophoresis, and Enzyme Activity Assay of the Luciferase Enzyme from Fireflies

Safety and Hazards   35
Introduction   35
Technical Review and Equipment Highlights   36
Experiment 4   41
Results and Discussion   42
Problem Set   43
References   43
Additional Materials   43

5. Hexokinase and G6PDH Catalyzed Reactions of Glucose Measurement

Safety and Hazards   45
Introduction   45
Technical Review   49
Equipment Highlight   54
Experiment 5   54
Results and Discussion   56
Problem Set   56
References   56

6. Polymerase Chain Reaction (PCR)

Safety and Hazards   59
Introduction   59
Technical Review   62
Experiment 6   64
Results and Discussion   65
Problem Set   66
Reference   66

7. Investigating Protein:Nucleic Acid Interactions by Electrophoretic Mobility Shift Assay (EMSA)

Safety and Hazards   67
Introduction   67
Equipment Highlights   69
Experiment 7   69
Results and Discussion   71
Problem Set   71
References   72

8. Qualitative Analysis of the Degradation of RNA via Ribonuclease A versus B

Safety and Hazards	73
Introduction	73
Experiment 8	74
Results and Discussion	75
Problem Set	76
References	77

9. Preparation of a Fluorescently Labeled Liposome and Its Analysis by Fluorescence Microscopy

Safety and Hazards	79
Introduction	79
Experiment 9	82
Results and Discussion	84
Problem Set	84
References	85

10. Studying Cell-like Structures with Liposome, DNA, and Protein

Safety and Hazards	87
Introduction	87
Experiment 10	88
Methods and Procedures	88
Results and Discussion	90
Additional Experiment	90
References	91

Index   **93**

Introduction

Biomolecules make up all the subcellular structures of cells and tissues. They can be classified into four main groups: (1) proteins, (2) nucleic acids (DNA and RNA), (3) lipids, and (4) carbohydrates. These groups are derived from the six primary bioelements (carbon, hydrogen, oxygen, nitrogen, phosphorous, and sulfur), which chemically bond together to form the diversity of structures that make up the four main classes of biomolecules. The importance of biomolecules and their interactions to the control and regulation of life cannot be overstated. For example, lipids, proteins, and carbohydrates organize into the structure that creates the boundary of the cell, namely the plasma membrane. In the nucleus, DNA is organized into chromosomes, which are primarily protein:nucleic acid higher-order structures. While there are many examples of biomolecules interacting together, one of the most beautiful and elegant is the ribosome, which is composed of multiple proteins and RNA molecules, and is organized into a structure that controls the translation of mRNA into every protein in the cell or secreted outside the cell.

In this laboratory textbook, the goal is for students to gain hands-on experience with these biomolecules, experimenting with them in a way that will give them a feel for how to work with proteins, nucleic acids, carbohydrates, and lipids and how they behave and interact with each other. The experiments become gradually more complex and move more toward reactions and interactions of these molecules with each other as the young investigators grow in experience and gain in confidence. Chapter 1 provides some basic rules regarding how students should conduct themselves in the laboratory to emphasize safety and good techniques. Then throughout the semester, students start out by doing a simple experiment designed to review basic techniques used in the biomolecular science laboratory, and then building on that, they learn to prepare a few buffers commonly used in molecular biology. In the following chapters, students investigate the four types of biomolecules and their building blocks; they gradually move to experiments designed to illustrate some typical biological chemistry and reactions and then finally model the various structures such as the chromosome or cell membrane.

Specifically in the first lab, students start out by refreshing basic skills, techniques, and mathematical operations required in the biomolecular science laboratory. Thus, in the first experiment, students review the basic techniques of liquid transfer, such as how to use the micropipette and how to choose the appropriate tools for transferring a certain volume. We also show students how to properly use an analytical balance. Then students conduct an experiment to confirm the density of water, calculate the molarity of a simple sodium chloride solution, and determine its weight percentage (%mass/volume). Using these determinations, students analyze the data for its variation and standard deviation and perform a statistical test.

In the second lab, students learn how to prepare biomolecular buffers appropriately. Commonly used buffers in the biomolecular science lab are prepared. These buffers include phosphate buffered saline (PBS), Tris-EDTA (TE), and SDS-Page gel running buffer (TGS). Each component of the buffer—for example, sodium phosphate (Na_2HPO_4) in PBS—is prepared at a specific molarity or concentration. Students check and double-check that calculations are correct so that buffers are made at the correct concentration in units of molar (M) or millimolar (mM). They adjust the pH of buffers such that they achieve the correct final pH for each buffer, calibrating and using the micro-pH meter appropriately with the correct standards. Each buffer must be prepared accurately because students will use them in subsequent laboratory experiments.

In the third lab and every lab thereafter, students begin explorations on biomolecules and their interactions. The third lab has to do with amino acids, which are the building blocks of proteins. Each amino acid is a unique biochemical with characteristic physico-chemical properties, which students test based on their solubility in various common solvents (water, acetonitrile, ethanol, etc.) used in biomolecular science. The amino acids' mobility upon thin layer chromatography (TLC) and hence their interaction between solvent and the molecules coated on the TLC plate are investigated.

In the fourth lab, students purify a protein of great interest to cell and molecular biology, the enzyme Luciferase (*Luc*), produced in the tails of fireflies, allowing them to flicker and light up in the summertime. The Luciferase protein is purified using standard precipitation, centrifugation, and filtration techniques. The protein is then analyzed for purity, and its activity is determined using a commercially available Luciferase assay kit. To produce light, *Luc* enzyme interacts with magnesium (Mg^{2+}) and ATP, oxidizing "Luciferin" substrate in its reaction. Students also take a portion of the purified protein and separate it on an SDS-PAGE gel to see how many kinds of proteins are present in the sample and confirm the presence of a protein at the correct molecular weight. Finally, they calculate the enzyme's specific activity and compare it to values in the literature.

In the fifth lab, students have a chance to learn how glucose concentration can be accurately determined by applying enzymatic reactions that are very similar to those in the human body. They use two enzymes, hexokinase and G6PDH, to catalyze the reaction, and then learn how to use standard solutions and a spectrophotometer to determine glucose concentrations in the samples. The reactions to be used here are typical of those in classic metabolic pathways such as glycolysis; they can also demonstrate to students how cofactors such as NADH are required in enzymatic reactions and indeed can be measured to monitor the progress of reactions. Finally, students analyze their data and make their calculations based on derivation of a standard curve they generate, a very common practice in bioanalytical science.

In the sixth lab, students amplify a gene, Capsid component (NC_001416) from Lambda phage DNA by polymerase chain reaction (PCR). Students analyze the designed primer sequences using the tools available from the National Center for Biotechnology Information (NCBI) website to predict the sequence of the DNA fragment that will be amplified by PCR. After their PCR reaction is complete, they will also analyze the PCR product on an agarose gel and compare the DNA size with their predicted number. Students refresh their background of this hallmark technique in molecular biology and put it into practice, a technique that involves many important biomolecular interactions and the interacting dynamics in different temperatures. These interactions include but aren't limited to the base pairing of the primers to the complementary sequences within the target DNA; the polymerase to the template; and its associated factors, ATP, magnesium, and deoxytrinucleotides (dNTPs).

In the seventh lab, students study interactions between a well-characterized DNA-binding protein, protamine, with the DNA fragment produced by the PCR reaction in the previous lab. Protamine is a short, highly cationic protein that is very similar in its sequence and function to histone proteins, which condense and protect the DNA in chromosomes. Protein:DNA interactions are extremely important in regulating gene expression and in controlling such processes as replication and transcription. In this lab, protein:DNA interactions are detected directly and confirmed by a technique called electrophoretic mobility shift assay (EMSA), which is used commonly in biochemistry and molecular biology. In EMSA, when proteins such as protamine bind to DNA, they cause the DNA's migration to change when it is electrophoresed through an agarose gel visualized by ethidium bromide staining. Here students also use common biochemical reagents (urea, ethanol, SDS) to disrupt the protein:DNA interaction and thus to find out what kinds of molecular interactions (hydrogen-bonding, ionic, or van der Waals interactions) are important in stabilizing protein:DNA interactions.

In the eighth lab, students analyze the kinetics of RNA degradation by RNase. There has been an upsurge of interest in RNA as being the "hidden gem" in the era of genomics and proteomics. RNA is more difficult to work with but holds promise for many discoveries, so in this straightforward lab, students are exposed to some basic techniques for RNA analysis and quantification. Here, they compare RNase A to RNase B, the glycosylated form of the enzyme. They also analyze differences in activity patterns and kinetics by using standard agarose gel electrophoresis, similarly to that done previously but now as applied to RNA stability.

In the ninth lab, students produce liposomes, which are essentially stripped-down empty cells containing a naked membrane. The liposomes contain a 1:1 mol:mol ratio of triaclylglyceride (tripalmitin) and cholesterol. Tripalmitin contains a C16 chain length, which is the most common and only fatty acid synthesized *de novo* in humans, whereas cholesterol is a lipid present in almost all cell membranes of animals, establishing proper membrane permeability and fluidity. A small percentage of fluorescently tagged cholesterol is incorporated into the liposome, thus allowing students to visualize it directly by using fluorescence microscopy.

Finally, in the tenth lab experiment, students build on the previous lab and incorporate DNA within the liposome, where the DNA can be condensed by protamine and then encapsulated to produce what are known in the literature as lipid-protamine-DNA (LPD) particles, which have been shown to be quite useful for gene delivery and gene therapy. In some senses, these LPDs can be thought of as very simple "artificial cells." The DNA can be stained by DAPI, and we have shown the protamine can be similarly stained, as published in *Nanomedicine* (2012). As in Lab 9, students will be able to observe these structures through the use of fluorescence microscopy.

In conclusion, every single experiment builds on itself, whereby students start out with water—undoubtedly the most important biomolecule for life—and review basic lab techniques. They learn how to prepare buffers appropriately and then investigate amino acids, which make up all the proteins in the human body. After they understand these properties, students perform a quick protein purification on Luciferase, an enzyme widely used in cell and molecular biology. When they begin to understand proteins better, students then investigate reactions involving metabolic enzyme catalyzing carbohydrates and PCR reactions to synthesize DNA fragments. Then they employ agarose gel electrophoresis and electrophoretic mobility shift assay for protein:DNA interaction and RNA degradation by RNase. Finally, they produce a liposome membrane, and utilizing a biomolecular conjugate (fluorescently labeled cholesterol), they visualize the structure through fluorescence microscopy and incorporate DNA and protein within it to produce a simple "artificial cell."

Answers for the Problem Sets at the end of each Experiment can be found on the companion website for this book: http://booksite.elsevier.com/9780128009697

About the Authors

ROBERT K. DELONG

Robert K. Delong's PhD education at Johns Hopkins and post-doctoral training at the University of North Carolina-Chapel Hill in biochemistry, biophysics, molecular biology and pharmacology provided an important foundation and background for this work. The early ideas for this book were clearly also related to my two industrial positions afterwards, first at a company best known as GeneMedicine and then at Powderject Vaccines. My groups were responsible for transferring formulations and processes from research into development and we performed many experiments on the biomaterials which feature in this book such as liposomes, proteins such as protamine and others, and many different types of DNA, oligonucleotides, plasmids and other gene constructs and expression vectors, etc. But it wasn't until I joined the faculty at Missouri State University and was asked to teach the biomolecular interactions class as part of the cell and molecular biology program that the unique need for such a book became clear. Senior faculty at the time Dr. Michael Craig and Dr. Chris Field (now faculty emeritus) suggested developing a series of unique experiments where the students could really get their hands dirty and learn by doing with these molecules. I remember our early phone conversations where they tried to get me to really think about what kinds of fundamental techniques, equipment and experiences should students have in this laboratory class. Over the semesters as each different cohort of students ran the experiments we tried to iteratively improve them, indeed involving a few honors students in optimizing one or two of the labs, and then as Dr. Zhou came on board she really helped to refine them adding a couple of really great new and exciting experiments. Although I always say "I am a scientist who teaches and not the other way around," having been very fortunate to have had my research involving anti-cancer RNA nanoconjugates funded by the National Cancer Institute now going on five years I can also say that the research I do with collaborators when it comes to RNA nanotechnology and therapeutics has also influenced this book somewhat. These days I am reviewing more and more research grants, and science funding agencies such as the National Science Foundation and the National Institute of Health are increasingly calling for more "translational science." In today's competitive and fast paced world it is thus not enough to simply know the important theories in science, or even the fundamental chemistry of each different biomolecule, but what is even more important, is to know how to apply this knowledge. It is here where we hope this book will serve the next generations of young investigators in biochemistry, molecular biology and physiology and biotechnology well. And to these students I say, be safe and follow your own path. It is my hope that you will pursue a career somehow related to science, or if not, if you will at least use its guiding principles in whatever career you choose. Having practiced as a scientist now going on 25 years, one thing I can honestly say it has taught me is humility and it has instilled within me a deep rooted belief in the beauty of our world. So in closing, no matter what you end up doing to make the world a better place, delve deeply into what interests you and don't get discouraged, tomorrow is a new day filled with all kinds of possibility and wonder.

QIONGQIONG ZHOU

Dr. Qiongqiong Zhou is an assistant professor at Missouri State University, where she studies cytoskeleton dynamics in her lab and teaches several core courses related to biomedical sciences. Prior to that, she was a research fellow in Douglas Robinson's lab at Johns Hopkins University and Ulrike Eggert's lab at Harvard Medical School.

Dr. Zhou graduated with a major of life sciences in 2002 from Fudan University in Shanghai, China. Then, she earned her Doctoral degree from University of Southern California, where she studies neurodegenerative diseases with Dr. Enrique Cadenas.

With a passion for undergraduate education, Dr. Zhou has led the lab of Biomolecular Interactions for more than 6 semesters using and continuously improving this lab manual. She has the most direct laboratory experience of these experiments with students.

Master Materials List

EXPERIMENT 1

Equipment

- Analytical balance
- Micropipettes (P20, P200, P1000)

Materials

- ddH_2O
- 12X Eppendorf microcentrifuge tubes (1.5 mL)
- Sodium chloride (NaCl) solution (1 M)

EXPERIMENT 2

Equipment

- Analytical balance
- Micropipettes (P20, P200, P1000)
- pH meter and calibration standards

Materials

- 25 and 100 mL graduated cylinders and volumetric flask for adjusting final volume
- 200 mL beakers
- Solutions: concentrated phosphoric acid, concentrated acetic acid, HCl (5 M) and NaOH (5 M) for pH adjustment, EDTA (0.5 M), SDS (10%)
- Chemicals: Tris-base, KCl, NaCl Na_2HPO_4, KH_2PO_4, Glycine

EXPERIMENT 3

Equipment

- Micropipettes (P1000, P200, P20)
- Drying oven (optional) or blow dryer

Materials

- Amino acids (Leu, Pro, Lys, Cys, plus an unknown amino acids mixture)
- 500 mL beaker
- Parafilm
- Premade ninhydrin staining solution (0.2% ninhydrin in ethanol, e.g., Sigma Ninhydrin spray reagent, Cat# N1286)
- Silica TLC plates (e.g., Millipore classical TLC plates, Cat# 105748) cut into size 6 × 8 cm
- ddH_2O
- Methanol (MeOH)
- Acetonitrile (ACN)
- Isopropanol (Ipr-OH)
- Ethanol (ETOH)
- 1 M acetic acid (CH_3COOH)
- 96-well plates, clear bottom
- Capillaries (optional, e.g., Fisher Cat# 22-260-943)
- Weighing spatula or forceps

EXPERIMENT 4

Materials, Equipment, and Procedures

- Ice cold acetone (reagent grade)
- Bench-top centrifuge
- Microcentrifuge tubes, plastic mortar and pestle
- Firefly tail and head to be used as control (1 per bench group, ½ tail or head per group)
- Luciferin (substrate for activity assay)
- Pipetteman and tips
- Sonicator
- PBS buffer substituted with 1% SDS (lysis buffer), PBS (suspension buffer)
- Zinc oxide nanoparticles (Sigma–Aldrich, Cat# 544906-10G)
- Gel electrophoresis supplies (prepoured gradient SDS/PAGE gel; 10X running buffer; loading buffer and power supply; prestained molecular weight ladder; Coomassie blue gel stain)
- Microtiter plate reader or simple luminometer (BMG or molecular devices)
- Enzyme buffer-luciferase reagent (Promega)
- Gel documentation station
- Protein staining buffer–GelCode Blue Thermoscientific Prod# 24592

EXPERIMENT 5

Equipment

- Microplate reader suitable for measuring absorbance at 340 nm
- Computer with Microsoft Office

Materials

- 96-well plates, clear bottom
- Glucose assay kit (sigma GAHK-20), contains glucose standard (1 mg/mL), enzymatic mixture of hexokinase, G6PDH, NAD, and ATP
- Micropipettes and tips
- Samples (Soda/juice, cell culture medium)

EXPERIMENT 6

Equipment

- Thermal cycler machine
- Micropipettes (P20, P200, P1000)
- Power supply
- Gel apparatus
- Microwave
- Gel documentation system

Materials

- PCR tubes
- 2X PCR master mix
- Primers:
  5 end primer (Sal I site):
  AAAAAAGTCGAC **ATGTCGATGTACACAACCGCCC**
  3 end primer (Not I site):
  AAAAAAGCGGCCGC **TTACGCCAGTTGTACGGACAC**
  (The underlined sequences are the template complementary sequences.)
- Lambda phage DNA as DNA template
- Nuclease free water
- Agarose powder
- TAE buffer
- Loading dye
- DNA ladder
- 250 mL flask
- 100 mL cylinder

EXPERIMENT 7

Materials and Equipment

- Horizontal gel electrophoresis apparatus and power supply
- Gel documentation station (or UV lamp, protective shield, and camera)
- Electrophoresis-grade low-melting-temperature agarose
- Protamine sulfate, Sigma P3369-10G (2 mg/mL dissolved fresh in water)
- Plasmid DNA (for example, Aldevron gWiz GFP Cat# 5006)
- Microcentrifuge Eppendorf tubes
- 1X TAE buffer (40 mM Tris-acetate, 1 mM EDTA, pH 8.3)
- 5X Loading buffer stock (40% sucrose)
- Loading buffer with tracking dye (0.1% Bromophenol blue [BPB] + 0.1% Xylene cyanol)
- Ethidium bromide (1 mg/mL solution premade by your instructor)

EXPERIMENT 8

Materials and Methods

- Torula Yeast RNA (Sigma–Aldrich)
- DEPC-treated RNase-free ddH$_2$O
- Bovine pancreas RNase A and B (Sigma–Aldrich)
- 1X TAE buffer
- Low-melting gel-grade agarose
- Gel electrophoresis equipment and documentation station with UV lamp and imager
- Bromophenol blue loading buffer (40% sucrose)

EXPERIMENT 9

Materials and Equipment

- Tripalmitin
- Cholesterol
- Chloroform (CHCl$_3$)
- Clean round-bottom (RB) flask
- Sterile PBS buffer
- Fluorescent microscope, slides, and cover slips
- Sonicator

- Fluorescent cholesterol—NBD cholesterol (Invitrogen/Molecular Probes N1148) or Bodipy cholesterol (Molecular Probes 5421563C11)
- Rotary evaporator with vacuum line (optional)
- Sterile agar plates for microbial growth and incubator
- Analytical balance (calibrated and certified prior to use)
- General lab equipment (pipettemen, tips, Eppendorf microfuge tubes, small weigh boats and paper, clean spatula, and glassware)

EXPERIMENT 10

Materials and Equipment

- Cholesterol
- Dioleoylphosphatidylcholine (DOPC) [Sigma–Aldrich or Avanti polar lipids]
- Chloroform ($CHCl_3$)
- Clean round-bottom (RB) flask
- Sterile PBS buffer
- Fluorescent microscope, slides, and cover slips
- Sonicator
- Fluorescent cholesterol—Bodipy cholesterol (Molecular probes 5421563C11)
- Rotary evaporator with vacuum line (optional)
- DAPI stain
- Analytical balance (calibrated and certified prior to use)
- General lab equipment (pipettemen, tips, Eppendorf microfuge tubes, small weigh boats and paper, clean spatula, and glassware)

Format of the Lab Notebook and Formal Lab Report

LAB NOTEBOOK

A laboratory notebook is a permanent record of an experiment's goals and outcomes, including all steps taken and observations made. Students must make original entries for each experiment in their laboratory research notebook. A suggested format is outlined here:

(1) Date

(2) Experimenter's Name

(3) Title

(4) Safety: It is important that you understand the risks involved with the chemicals used within the laboratory and how to safely handle and work with them. This section should contain pertinent information that is vital to your health, safety, and well-being.

(5) Purpose: The purpose should include one brief and concise sentence about why you are conducting the experiment and what you hope to accomplish.

(6) Introduction: The introduction should provide very brief background pertinent to the current experiment. Any equation, theory, or law to be put in practice should be listed. If there are any key publications prior to this that provide background information, this material should be referenced from the lab text as well as any other outside academic sources.

(7) Materials: List all the reagents and chemicals that this lab will use, including name, concentration, and pH. Pay extra attention to the safety and handling guidelines for some chemicals and reagents if needed (for example, how to dispose of EtBr, keeping an enzyme solution on ice when working with it on the bench to preserve its activity, etc.).

(8) Methods: This section should include a detailed outline or flowchart of the experiment procedure to be conducted. You may find it helpful to make additional notes on the margins in this section as you work through the day's lab. Also, note any changes to the protocol and procedures actually performed during the lab.

(9) Results and Data Interpretation: Record your data in this section. Data can be qualitative as well as quantitative; all observations should be recorded. This information will be especially helpful if you are asked to go back and troubleshoot where your experiment may have gone wrong. Try your best to organize the data into easy-to-follow tables, figures, or graphics. You should also analyze and interpret the data in this section. Interpretation of data is crucial; it is a time of analytical reflection. Did the data support the goal you hoped to accomplish? If so, how? If not, why not? Do these unexpected results lead to additional questions? What might you do differently if you were to repeat this experiment?

(10) Conclusions and Discussion: Include one paragraph, at most, summarizing your findings and putting it into context of what is in the literature in the area as appropriate. Here, you may also discuss what could possibly cause your data to be different than expected and how to improve your experiment if you are going to conduct it again.

(11) References: Include a reference section at the end. This section cites manuscripts, web links, or any other sources referred to in your write-up. Include *at least* the main author, journal title, volume, page numbers, and year of publication.

FORMAT OF A FORMAL LAB REPORT

A formal lab report follows similar outlines as your lab notebook, except that it has to be written with formal language.

(1) Title: Include a short but descriptive title.

(2) Authors: List all authors.

(3) Abstract: Add a brief summary of the background, the experiments performed, the results obtained, and the conclusion in approximately 200–300 words.

(4) Introduction: Introduce readers to the topic of the work; give background on the subject, the problems you are trying to solve, and the significance of this work.

(5) Materials and Methods: Use descriptive language. Instead of listing steps to describe the methods, you can imbed the materials in the methods.

(6) Results: Use descriptive language to explain your results. What is the purpose of this particular experiment, what method did you use, where is the data shown, and how do you interpret the data? Insert your figures or tables in this section and refer to them in the result context.

(7) Conclusions and Discussion: Make your conclusion based on all your results, the problems you had during this work, the possibilities that may solve those problems, and the significance of your work and the future directions that this work may lead to.

(8) References

Lab Safety and Policies

SAFETY

SAFETY IS THE #1 CONCERN IN THE LABORATORY AT ALL TIMES. You must bear the safety of yourself, your fellow students, and coworkers in this laboratory; other laboratory classes; and the research lab as being the most important thing at all times, no exceptions. To illustrate the importance of safety, I would like to share a story from my own personal experience.

One day while I was in graduate school, I was innocently passing through the laboratory carrying a bottle of solvent when an explosion went off in the chemical hood across the lab from me. My colleague working in the lab had left the sash of the hood open, and although he was wearing glasses, which thankfully protected his eyes, exploding glass shards lodged in his arms and torso, and he went into shock temporarily. My colleague's injury could have been avoided if the hood had been properly closed and my fellow scientist had followed proper safety procedures. Twenty years later he has since had no such accidents, and suffice it to say, this experience taught him a very valuable lesson. *That is, safety is the paramount consideration any time you are working in the lab.*

Safety equipment: In the preceding example, the accident could have been avoided if the chemical reaction had been set up in the hood, and the sash or sliding door in front of the chemical hood had been closed. All chemicals capable of catching on fire or exploding should always be stored in a closed chemical hood or chemical safety cabinet. Personal protection equipment (PPE) should be worn in the lab at all times, including safety glasses or goggles, and a lab coat (or at the very minimum, clothing that covers all exposed skin, legs, arms, and torso). Shoes or sneakers that cover the entire foot should also be worn to protect the foot in the event of broken glass or a spill. You should familiarize yourself with safety equipment in and nearby the laboratory, in case you ever need to use it. When I fly on an airplane and the flight attendant reviews the safety procedures, I pull out the safety card and review it *every time*. In an emergency, you may have only a second or two to make a split-second decision. In those seconds may rest the fate of everybody involved. There are many examples of people performing heroic procedures, but only if they know where the safety equipment is and how to use it. In addition to the chemical hood and safety cabinet, this includes an eyewash station and a safety shower.

The eyewash station should be used freely and liberally without worry of water damage any time a student has splashed a chemical or reagent into his or her eyes. The safety shower should be used in cases in which a spill involves a large portion of someone's body or any time any part of a person's clothing or body catches on fire. A fire extinguisher may also be available nearby the lab you are working in; if so, as with all safety equipment, be familiar with where it is and how to use it.

Personal responsibility and behavior: Science is fun, even magical and awe-inspiring at times. A great deal of joy can be had by learning to practice it and by branching into an area or specialty that interests you. That said, although there are many venues in college or at university where you should blow off steam, the lab is no place to fool around or engage in horseplay. With dangerous chemicals, sharp objects, and expensive equipment often in a fairly tight space, the science laboratory is a certain "recipe for disaster" if you don't conduct yourself safely and responsibly. Of course, there should be no food or drink, smoking, or consumption of alcoholic beverages in the laboratory. You should remove your jacket and backpack from the working area and remove any other clutter to avoid tripping.

Remember, the laboratory is the place where your instructor can work closely with you and learn what makes you tick; it also is the place where you get to show him or her what you are capable of. Therefore, your lab instructor is the one most likely to write a recommendation that nets or lands you that job or admission to graduate school later. Think about behaving in the lab the way that you anticipate acting when you graduate from college and take your first job. People learn by doing, and the experiments in this book were designed with that approach in mind: to get you working with your hands with the very molecules that control life at the subcellular level. Take advantage of that opportunity and work safely and responsibly. After your lab session is a good time to work out, go for a bike ride, swim, play a game of basketball or tennis, or do whatever you do to relax. So don't mess around while in the lab, work hard, try your best to engage and stay focused on learning, and your work in the lab will be an incredibly rewarding experience on many levels.

Material Safety Data Sheets (MSDS): Another very important consideration to working in the laboratory is to always familiarize yourself with these MSDS sheets *before* working with the chemicals and reagents in any experiment. MSDS sheets contain important information for what to do when a chemical is accidentally spilled or the correct way to dispose of it. If you are following appropriate procedures, working with chemicals in the chemical hood, and wearing protective clothing and eyewear, there should be no cause for concern. However, if an accident does occur, these MSDS forms provide important information for what to do, for example, in case of a chemical spill on the skin or otherwise. MSDS sheets are available on the Internet (for example, chemexper.com or chemfinder.com). Per university policy, copies are maintained in the lab in the MSDS master binder in case of such spills and accidents. You should always consult these MSDS forms before working with chemicals and reagents in the lab to be aware of any potential dangers and hazards.

Working safely in the lab, disposal, and clean-up: The laboratory experiments in this book were designed with student safety in mind. However, you should be aware of specific reagents used in the various labs and will want to take extra precaution when working with them. For example, in the buffer lab (Lab 2), you need to carefully pipette small amounts of strong acids and bases into the buffer in order to adjust the pH in the correct range. Acids and bases are caustic; therefore, you should wear gloves and work in the hood for these steps. In the amino acids lab (Lab 3) and liposome lab (Lab 9), you work with several solvents that are quite toxic and should never be ingested or spilled onto the skin because their absorption into the bloodstream and body would be very dangerous. These include acetonitrile (a derivative of the potent poison cyanide), chloroform, and several others. In addition, several of the stains used to visualize biomolecules in these laboratory experiments, such as ethidium bromide (EtBr) or ninhydrin, are also mutagenic or carcinogenic (cancer-causing). You should perform these staining steps while wearing gloves ideally in the hood to avoid any possible spill onto the lab bench, which might contaminate the general working area and expose other students. Care should be taken to dispose of these toxic agents appropriately, and your instructor will provide details on this in the pre-lab lecture. Finally, some of the equipment used in the lab can be dangerous unless you operate it properly. This includes but may not be limited to centrifuges, which must be balanced and the lid closed before starting a spin, and electrophoresis units and power supplies, which represent an extreme electrical current shock hazard and should be turned off until properly set up and the apparatus set up and closed properly. Remember to never pour anything down lab drains except water (unless otherwise specified). Properly label and cap all reagents and waste containers. Dispose of waste in appropriate containers. (For example, "Halogenated," "Carcinogenic," "Organic Waste," and "Mutagenic" are examples of labels on waste containers.) At the conclusion of each lab, properly store equipment and reagents; also thoroughly clean glassware and lab bench surfaces. Not doing so may result in points being deducted from your lab write-ups.

Safety contract: A signed safety contract for all students in the lab must be on file with your instructor prior to conducting any experiment (see the following section). This represents a student's pledge that you will abide by and follow the instructions written here for safe conduct in the laboratory, that you will familiarize yourself with the safety equipment and appropriate MSDS sheets before each lab, that you will come to the lab properly clothed and prepared, and that you will work safely while performing the lab exercises.

SIGNED SAFETY CONTRACT

I hereby acknowledge that I have read the "Lab Safety and Policies"section of the *Introductory Experiments on Biomolecules and Their Interactions* text in its entirety and agree to abide by the policies and procedures set forth in this document. Furthermore, I have read, understand, and agree to abide by all of the safety information.

I understand that failure to follow these procedures and conduct myself safely in the laboratory may result in my being asked to leave the laboratory without question for the safety of myself and others, and the forfeiture of any associated points for that lab. Repeat offenders who are asked to leave the lab more than once for not following these procedures and for endangering themselves or others while conducting the lab will be dropped from the lab, and hence the course.

Full Name (printed)　　　　　Signature　　　　　Date

(Please complete this form, remove it from the manual, and give it to your instructor. This is a contract. Your instructor will maintain it in his or her lab files as a record. Without this signed form, you will not be allowed to participate in the lab. Please note also that from time to time it may be necessary to change or modify the protocol, experiment, or even day of the lab to take advantage of updated information.)

1

An Introduction to Basic Math and Operations in the Biomolecular Laboratory

SAFETY AND HAZARDS

While there are no particular health and safety concerns in particular for this lab, you should refamiliarize yourself with the safety equipment in the lab (chemical hood, eyewash station, shower, etc.) and review the proper handling, storage, and disposal of solvents and other reagents and chemicals to be used in future labs.

INTRODUCTION

Water (H_2O) is perhaps the most important biological molecule (biomolecule). It dissolves many other biomolecules and is the major component in biological fluids such as blood, interstitial fluid, and the fluid within cells (cytosol). It also helps regulate pH and temperature, reacts and interacts with many other biomolecules, forms hydrogen bonds to itself and other biomolecules, and helps maintain biological structures, etc. Indeed, it could be said that all living organisms have a critical and ongoing requirement for water. Most reactions and interactions you seek to emulate or manipulate in the biomolecular laboratory require water and a few key biomolecules dissolved in a water-based solution. Almost always these interactions and reactions you explore in laboratory experiments require a certain specific concentration of these key biomolecules dissolved in water. Since a biomolecule's concentration is dependent on two factors—mass (M) of the solute (the molecule to be dissolved) and the final volume (V) of the solution—it becomes very important for you to understand how to appropriately dispense or aliquot accurate volumes

1

and to be able to weigh out a specific mass of the biomolecule. There are many examples, but in these experiments, this step can include calculating the concentration of buffer, ions, and H^+ (pH) and controlling the concentration of enzyme and substrate and any key ions or cofactors in enzyme reactions. So Lab 1 first reviews the basic mathematical principles that underlie being able to calculate concentrations correctly and the equipment most frequently used to prepare such biological solutions, such as mass balance for weighing the mass of a biological compound and various types of pipettes used commonly in the lab. Finally, you conduct an experiment, essentially weighing volumes of water dispensed in the microliter (μL) range or those containing sodium chloride (NaCl), checking correct use of basic math and statistics used commonly in the biomolecular lab.

TECHNICAL REVIEW

This section reviews basic mathematical operations, calculations, and conversions you typically encounter in the biomolecular laboratory. Typical basic math operations involve scientific notation, significant figures, accuracy, and precision and some simple statistical analyses of your data (mean, range, standard deviation, and student's t-test). Review of these math concepts follows.

Scientific notation and conversions: In science, you often use scientific notation in your notebooks to record a number or data point(s). For example, if you weighed a sample and it was 1.45 mg (and you wanted to convert this sample into grams), you would first need to know or look up the conversion factor (1 g = 1000 mg). Then you perform the math: [1.45 mg × (1 g/1000 mg)] = 0.00145 g. In scientific notation, this would be 1.45×10^{-3} g. Table 1-1 summarizes common units of length, mass, volume, and molarity and their conversion factors.

Significant figures: Generally, analytical balances and pH meters are accurate only to the tenth (0.1–0.9) or at best hundredth place (0.01–0.09). So, in the preceding example, if the balance were accurate only to the tenth place, you would round up and record the data in your notebook as

$$1.5 \times 10^{-3} \text{ g}$$

Accuracy and precision: Accuracy and precision are pictorially represented in Figure 1-1.

For example, if you weighed three samples and the data were #1 = 29.1 g, #2 = 29.4 g, and #3 = 29.7 g, you might conclude that your measurements are precise, as they are close together. However, if you had a true standard weight of 30 g (weighed on the balance) and found that the balance read 20 g, then you would realize that the balance, though precise, was not accurate (and would need to be recalibrated).

TABLE 1-1 Common Units of Length, Mass, Volume, and Molarity and Their Conversion Factors

Length unit	Mass unit	Volume unit	Molarity unit
1 Km = 1000 m	1 Kg = 1000 g	1 L = 1000 mL	1 M = 1000 mM
1 m = 100 cm	1 g = 1000 mg	1 mL = 1000 μL	1 mM = 1000 μM
1 cm = 10 mm	1 mg = 1000 μg	1 μL = 1000 nL	1 μM = 1000 nM
1 mm = 1000 μm	1 μg = 1000 ng		1 nM = 1000 pM
1 μm = 1000 nm			

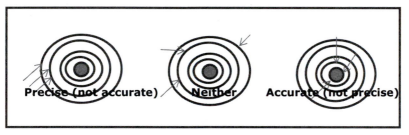

Precise (not accurate)   **Neither**   **Accurate (not precise)**

FIGURE 1-1   Accuracy and precision

Statistical analyses: Typically, most experiments are run in triplicate and repeated at least once if not twice (replicated). Good practice prior to publication of the result is for another scientist or investigator to independently repeat the result—or even better, another lab. For example, experiments are often replicated at a constant time, temperature, or concentration; sometimes these are the variables of interest in the experiment. In order to gain confidence in data and make comparisons between control groups and the corresponding experimental groups, scientists often determine the mean and range, and perform standard deviation or the student's t-test. This gives a quantifiable conclusion based on the experiment. It allows you to tell how different the two groups may be.

Data analysis software such as Microsoft Excel can help you analyze the data (calculate the mean, standard deviation, etc.), but first you should know these equations to give formulations to the software so that it knows how to analyze. This process is summarized briefly here.

Mean: The mean or average (sometimes denoted X) is simply calculated by summing up the individual data points and dividing them by the number of individual measurements taken. Using the preceding example in which you took three sample weight measurements (n = 3):

$$\textbf{Mean} = (29.1 \text{ g} + 29.4 \text{ g} + 29.7 \text{ g}) / 3 = 29.4 \text{ g}$$

Range: The range of this experiment is simply the difference between the smallest and largest measurement taken:

Range = Largest value in dataset − smallest value in dataset = 29.7 g − 29.1 g = 0.8 g (range)

Standard deviation: The standard deviation (S.D.) is a calculation that takes into account the mean, the sum of the differences between the mean $\Sigma(X_i - mean)$, and a number of measurements (n) that are taken. It is commonly used to determine if a set of measurements is significantly different from another control group. The equation to calculate standard deviation is

$$\mathbf{S.D.} = \sqrt{\left[\Sigma(X_i - mean)^2/n - 1\right]}$$

So from the previous example of sample weights,

$$\mathbf{S.D.} = \sqrt{\left\{\left[(29.7 - 29.4)^2 + (29.4 - 29.4)^2 + (29.1 - 29.4)^2\right]\Big/(n - 1)\right\}}$$

$$= \sqrt{\{[(0\ 09) + 0\ 0 + (0\ 09)] / (3 - 1)\}}$$

$$= \sqrt{\{[0.18/2]\}}$$

$$= 0.3$$

Thus, it is common to take the mean +/− S.D. (in this case 29.4 +/− 0.3).

The *student's t-test,* developed by W. S. Gossett (1876–1937), is typically used when dealing with problems associated with small sample size. It allows you to determine with some degree of confidence that the mean in your experiment (limited sample size) is reasonably close to the "true mean," or the two sets of data are really different. Following is an equation using the student's t-test to calculate *confidence intervals (CI),* where the CI is a measure of the reliability of an estimate, which usually gives a range based on a group of sample data, X_{mean} is the mean of the samples, and S_m is the standard error. Table 1-2 shows a selected list of t values at a certain confidence level and certain degree of freedom ($df = N - 1$). If the confidence level is set at 95%, then you may say with 95% confidence that the true value is located within ($X_{mean} + (t) * S_m$, $X_{mean} - (t) * S_m$), where the t value is selected according to the confidence level and *df* in Table 1-2.

$$CI = X_{mean} +/- (t) * S_m$$
$$S_m = S.D./\sqrt{n}$$

TABLE 1-2 A Selected List of t Values for Student's t-Test

t VALUES FOR VARIOUS VALUES OF df

α LEVEL TWO-TAILED TEST

	80%	90%	95%	98%	99%	99.80%	99.90%
	0.2	0.1	0.05	0.02	0.01	0.002	0.001

α LEVEL ONE-TAILED TEST

df	0.1	0.05	0.025	0.01	0.005	0.001	0.0005
1	3.078	6.314	12.706	31.821	63.657	318.313	636.589
2	1.886	2.92	4.303	6.965	9.925	22.327	31.589
3	1.638	2.353	3.182	4.541	5.841	10.215	12.924
4	1.533	2.132	2.776	3.747	4.604	7.173	8.61
5	1.476	2.015	2.571	3.365	4.032	5.893	6.869
6	1.44	1.943	2.447	3.143	3.707	5.208	5.959
7	1.415	1.895	2.365	2.998	3.499	4.785	5.408
8	1.397	1.86	2.306	2.896	3.355	4.501	5.041
9	1.383	1.833	2.262	2.821	3.25	4.297	4.781
10	1.372	1.812	2.228	2.764	3.169	4.144	4.587
11	1.363	1.796	2.201	2.718	3.106	4.025	4.437
12	1.356	1.782	2.179	2.681	3.055	3.93	4.318
∞	1.282	1.645	1.960	2.327	2.576	3.091	3.291

EQUIPMENT HIGHLIGHT

In keeping with the ability to control the volume of a biological solution or the mass of a reagent or biomolecule dissolved into it, for this experiment, you need to review the operation and functioning of the mass balance and pipettes prior to designing and conducting this lab.

Mass balance: Analytical balances are accurate and precise instruments used to measure masses. A typical analytical balance measures masses precisely up to 0.0001 g. Use these balances when you need high degrees of precision. But before weighing anything, you must make sure that the analytical balance is leveled. To level your balance, check the leveling bubble, which should appear in the center of the chamber. If the leveling bubble is not centered, you can adjust it by turning the balance's leveling screw. Once the balance is leveled, close all the chamber doors, make sure the weighing pan is not touching the draft ring, press the on/off button on the front of the balance, and then a row of zeros will appear. This indicates that the balance is zeroed and ready for use. If your weighing substances are liquid, powder, or granules, they must always be weighed using an appropriate weighing container, such as Eppendorf tubes, weighing paper, and weighing boat. Place the weighing container on the weighing pan and close the doors; then tare the container by pressing the tare button and wait for the zeros. This allows the mass of your sample to be read directly. Once the container is tared, add the substance to be weighed. Be careful not to spill chemicals on the balance. With the sample and its container sitting on the pan, close the chamber doors and read the display to find the mass of your sample. A typical mass balance is shown in Figure 1-2.

FIGURE 1-2   Mass balance instrument

Quantitative liquid transfer (pipetting): Liquid transfer is a very common technique used in many laboratory settings. The volume of the liquid in each experiment can vary quite a lot from several liters to a half microliter. Choosing the right measurement is crucial to meet the requirements of the amount and the accuracy in each experiment. Graduated cylinders or volumetric flasks are often used to measure volumes from 10 mL to 1 L, standard pipettes are used for volumes from 1 mL to 50 mL, and the smallest volumes from 0.1 μL to 5 mL are usually handled by micropipettes, also called *pipettemen.* Sample liquid transferring tools and their ranges are listed in Table 1-3.

Standard pipettes: A variety of different types of pipettes and bulbs are used commonly in scientific laboratories. They range from simple disposable Pasteur pipettes to pipette bulbs. For volumes in the 0.1 up to 25 or 50 mL range, volumetric, serological, or Mohr pipettes are often used. When sterility is required to guard against unwarranted cross-contamination or microbiological contamination, presealed and sterilized pipettes are necessary.

Micropipettes: The micropipettes are calibrated to dispense various volumes accurately. For example, a P1000 is accurate in the range of about 200 μL up to 1 mL (1000 μL). A P10 is generally accurate for dispensing volumes of 1–10 μL. These are used in conjunction with plastic pipette tips, which are often presterilized or pretreated for safe use with cells, RNA, or DNA work. A micropipette's plunger has three positions, as shown in Figure 1-3. In position 1, the pipette is at rest; position 2, corresponding to the set volume on the micropipette, is reached by pushing down on the

TABLE 1-3   Quantitative Liquid Transferring Tools

Tools		Capacity	Inaccuracy (+/− mL)
Graduate cylinders		10 mL–1 L	~1% of maximum volume
Serological pipettes	50 mL	50 mL	1 mL
	25 mL	50 mL	0.2 mL
	10 mL	10 mL	0.1 mL
	5 mL	5 mL	0.05 mL
	1 mL	1 mL	0.01 mL
		Range	**Inaccuracy (+/− μL)**
Micropipettes	P1000	200–1000 μL	40 μL
	P200	20–200 μL	20 μL
	P20	2–20 μL	0.4 μL
	P10	0.5–10 μL	0.2 μL

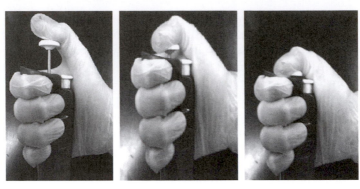

FIGURE 1-3   Proper operation of a micropipette

plunger until resistance is met. Position 3 is reached by pushing down further from position 2 to expel all the liquid from the tip.

EXPERIMENT 1

In this laboratory experiment, you put basic math concepts and high-lighted equipment (mass balance and micropipettes) into good use. The goals of this lab are twofold:

- To introduce you to various types of pipettes, pipetting techniques, and proper use of the analytical balances
- To introduce you to sample calculations and simple statistical methods

Equipment

- Analytical balance
- Micropipettes P20, P200, P1000

Materials

- ddH$_2$O
- 12X Eppendorf microcentrifuge tubes (1.5 mL)
- Sodium chloride (NaCL) solution (1 M)

METHODS AND PROCEDURE

1   Obtain 12 Eppendorf microcentrifuge tubes (1.5 mL).
2. Arrange them into four sets (A, B, C, D) and label each tube according to Table 1-4.

TABLE 1-4   Four Sets of Measurements

	A: H_2O 10 μL		B: H_2O 100 μL		C: H_2O 1 mL		D: NaCl 1 mL	
Repeat 1	A1		B1		C1		D1	
Repeat 2	A2		B2		C2		D2	
Repeat 3	A3		B3		C3		D3	
Average								
SD								

3. Place a labeled Eppendorf tube from step 2 into the center of the analytical balance; then press the tare or zero button until the digital display reads zero.
4. Choose a proper micropipette and add 10 μL, 100 μL, or 1 mL of water or 1 mL of NaCl solution to the tube. Place the tube back to the analytical balance and record the weight in Table 1-4. Repeat steps 2 and 3 until all the experiment sets are measured.
5. Record your data in Table 1-4; then calculate the average and SD in each set.

RESULTS AND DISCUSSION

The last section of your lab notebook write-up should be a brief summary and reflection of your data, what went well, what went differently than you thought, and any conclusions you can draw. To get started, consider writing about the following issues in this lab's Results and Discussion section:

1. In your calculations performed for the problem set, explain any trends or insights from your data.
2. Discuss your results. For example, given the density of water, what would you predict the weights of each should be? Explain any discrepancy you observe.
3. If you happened to use water contaminated with high concentrations of heavy metals, what would you predict would happen to your data? What other experimental error or equipment malfunctions could contribute to a lack of confidence in your data? Again, when you use water or make buffers with water, this is the reason you often use double-distilled, deionized water to avoid any unwanted contaminants or ions being introduced into the biomolecular interactions or reactions.
4. How could you improve this experiment?

PROBLEM SET

Be sure to consider significant digits and scientific notation:

1. Calculate the mean, range, and standard deviation of each data set (10 μL, 100 μL, 1 mL) from Table 1-4.
2. Perform a student's t-test on each data set to calculate the confidence limit of 95%.
3. What is the weight percent (% mass/volume) of the saturated sodium chloride?
4. What number of NaCl moles are dissolved in 1 mL of 1M NaCl? Convert this into mmoles and μmoles.
5. Choose the right tools to transfer the liquid.

2 μL,	5 mL,	350 μL,	2 L
800 mL	15 μL,	50 μL,	15 mL

Additional Materials

[1] Rodney Boyer, Biochemistry Laboratory—Modern Theory and Techniques, 2nd ed., 2006, Pearson Prentice Hall, 2012, Chapter 1.
[2] Jeffrey Paradis, Kristen Spotz, Hands On Chemistry, McGraw-Hill, 2006.
[3] John F. Robyt, Bernard J. White, Biochemical Techniques: Theory and Practice, Brooks-Cole Publishing Company, 1987.
[4] David S. Page, Principles of Biological Chemistry, Willard Grant Press, 1981, 1976.

Notes

_____

_____

_____

_____

_____

_____

_____

_____

_____

_____

2

Preparing Buffers at a Specific Molarity and pH

SAFETY AND HAZARDS

Small quantities of concentrated acids or bases (acetic acid, HCl, NaOH) are required to adjust the final pH of each buffer. *These should be used with extreme caution in the chemical safety hood and while wearing gloves* because they are caustic to skin and have volatile or strong, harmful vapors. Again, you must work with these under safe, controlled conditions and be very careful to avoid skin contact and not to inhale the vapors.

INTRODUCTION

Buffers are used commonly in biomolecular science. Most often they are used to dissolve biological molecules and maintain a constant pH range to protect their structure-function and avoid any acid or base-catalyzed degradation. A buffer is an aqueous solution containing a mixture of a weak acid and its conjugate base or a weak base conjugate acid. It is able to maintain the pH in a narrow range even when a small amount of strong acid or base is added to the buffer. Buffer agents are the primary components to make buffers, which function by sequestering protons (H^+) from solution and thus maintain the pH within a certain range. The H^+ binding affinity is determined by the equilibrium constant (K_b or K_a), which is specific to each particular buffer reagent. You can essentially think of buffers as molecular sponges that are able to sponge up H^+ at various concentrations.

Besides maintaining pH value, buffers have other applications, as is the case with phosphate buffered saline (PBS). In addition to dissolving proteins and other biomolecules, the osmolality and ion concentrations of PBS usually match those of the human body, so it is isotonic and nontoxic to cells. Also, PBS can be sterilized and used to wash and/or suspend or transfer cells. Another example is Tris-acetate-EDTA (TAE) buffer, which can be used to dissolve DNA or RNA (and hence may sometimes include DNase or RNase inhibitors). The Tris is the buffer that maintains the optimum pH for the biomolecule, and the EDTA component is used to scavenge any metal ions such as iron, mercury, etc., that might catalyze metal or radical-based cleavage of the biomolecules. TAE is often used as a gel running buffer for agarose-based gel electrophoresis. Similarly, Tris-glycine-SDS buffer is sometimes called running buffer for performing polyacrylamide gel electrophoresis (PAGE). Sodium dodecyl sulfate (SDS) is a detergent used to denature proteins during electrophoresis. The preceding three common buffers, their pH range, and a protocol for their preparation and some uses are summarized in Table 2-1. Often, some buffers are prepared as concentrated solutions (10X) that can then be diluted with double-distilled, deionized water (ddH$_2$O) prior to use.

TECHNICAL REVIEW

Molarity: Scientists use this term to mathematically describe concentration. It is often given in units of molarity (M), moles per liter; or in millimolar (mM), millimoles per liter. For example, if you need to prepare a stock solution of 500 mL of 1 M sodium chloride (NaCl), you can calculate this as shown in Equation 2-1:

$$Mass = C \times V \times Mw$$

where C = concentration, V = volume, and Mw = molecular weight. For your reference, consider the following molecular weights: Na = 23.4 g/mole, Cl = 35.5 g/mole, NaCl total M_W = 58.9 g/mole.

The next step is to set up the equation and be sure the units cancel out:

$$1\,(mole/L)*500\,mL*58.9\,g/mole = 29.5\,g \qquad \text{Eq. 2-1}$$

As shown in Equation 2-1, in this case, you would then dissolve 29.5 g in 500 mL to achieve a final concentration of 1 M. In this example, note that in terms of significant digits, the 1 molar (1 M) has only one significant digit; therefore, it would be okay to dissolve 30 g in 500 mL and express your final answer as 3×10^1 g. It is important to note that in making up aqueous solutions and buffers, you should always use double-distilled, deionized water (ddH$_2$O) to dissolve the reagent. Doing so ensures that there are no unwanted dissolved ions, which might otherwise potentially have adverse effects on the experiment or the biological macromolecules under study.

Dilutions: A stock solution is a concentrated solution that will be diluted to some lower concentration for actual use, called a working solution. To prepare a working solution from a stock solution, you first calculate a dilution factor. Using the earlier example (1 M NaCl), if you want 100 mL

TABLE 2-1 Common Buffers, Solutions, and Their pH Ranges

Buffers	pH range	Components	Final conc.	Volumes or mass	Uses
100 mL of TAE (10X)	8.0	Tris-base	400 mM	(____g)	To dissolve agarose for agarose gels and in the electrophoresis running buffer
		Acetic acid (99%)	200 mM	(____mL)	
		EDTA (0.5 M)	10 mM	(____mL)	
100 mL of PBS (1X)	7.2	NaCl		0.8 g	To dissolve biomolecules and for washing and resuspending cells and tissues (when sterile)
		Na$_2$HPO$_4$		0.12 g	
		KH$_2$PO$_4$		0.02 g	
		KCl		0.02 g	
100 mL of TGS (10X)	8.3	Tris-base	250 mM	(____g)	To run SDS-Page gels to separate proteins
		Glycine	1.92 M	(____g)	
		SDS (10%)	1%	(____mL)	

of physiological saline (150 mM NaCl), you would need to dilute the stock solution (1 M) by a factor of 1 M / 150 mM = 1000 / 150. The volume of the stock solution needed would be 100 mL / (1 M/ 150 mM) = 15 mL. You would then take 15 mL of the 1M NaCl stock measured out in a graduated cylinder and fill ddH$_2$O up to the 100 mL mark.

Buffer and the pH: Recall that pH is a simple mathematical equation that refers to the concentration of H$^+$ ions in solution, as shown in Equation 2-2:

$$pH = -\log [H^+] \qquad \text{Eq. 2-2}$$

Thus, an acid or a buffer's tendency to associate or disassociate H$^+$ is defined by its equilibrium constant (Ka). Ka is the equilibrium constant (K = [products]/[reactants]) that reflects whether the buffer is bound by H$^+$ or not:

$$HA \rightleftarrows H^+ + A^-, \text{ where } Ka = [H^+][A^-]/[HA$$

For acids, their conjugate bases, or buffers, the relationship between pH and pKa is called the Henderson–Hasselbach equation, as shown in Equation 2-3:

$$pH = pKa + \log \{[A^-]/[HA]\} \qquad \text{Eq. 2-3}$$

This equation essentially shows how much dissociation occurs at a given pH.

EQUIPMENT HIGHLIGHT

In this lab, you use a pH meter to adjust your buffer's pH value. A pH meter can sense pH by measuring the voltage difference between your sample supplying H^+ units versus a standard electrode often filled with saturated KCl or AgCl. Basically, the concentration of H^+ affects an electron exchange between these salts relative to the standard; hence, pH is sensed electronically. A typical pH meter and its basic components are shown in Figure 2-1.

Although the model and program in each pH meter may vary, you should always follow some common practice guidelines:

1. Always rinse the electrode with ddH_2O and gently dab away any excess fluid from the tip with a KimWipe before immersing it into a new solution or its storage solution.
2. Prior to measuring your sample, you should perform a two-point calibration using presupplied pH standards at 3.0 and 7.0 or at 7.0 and 10.0, which are the units below and above where your buffers need to be prepared. For example, if the solution you are making has a pH at 8.0, use pH 7.0 and 10.0 standards for two-point calibration; if the solution has a pH at 5.0, then you would use pH 3.0 and 7.0 standards for the calibration.
3. When measuring the pH of your solution, make sure that your solution is dissolved and mixed well, either by gently shaking the beaker or using a magnetic stir bar/stir plate combination. Immerse the electrode in the solution and wait a few minutes until the meter shows a constant number. If the number is higher than your desired pH value, add acid slowly to bring down the number, and vice versa. While adjusting the pH value with acid or base, always give some time for the meter to reflect the actual pH value. Do not add in a large amount of acid or base at once because doing so may overadjust the pH value in the buffer.

FIGURE 2-1 Typical pH meter, probe, and color standards (pH = 4, pH = 7, pH = 10)

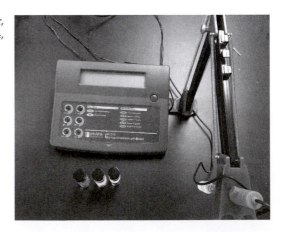

4. When finished, remember to turn off the pH meter and put the electrode back in its storage solution! Never leave the electrode in the air for a long time because drying out the electrode may damage it.

EXPERIMENT 2

Equipment

- Analytical balance
- Micropipettes (P20, P200, P1000)
- pH meter and calibration standards

Materials

- 25 and 100 mL graduated cylinders and volumetric flask for adjusting final volume
- 200 mL beakers
- Solutions: concentrated phosphoric acid, concentrated acetic acid, HCl (5 M) and NaOH (5 M) for pH adjustment, EDTA (0.5 M), SDS (10%)
- Chemicals: Tris-base, KCl, NaCl Na_2HPO_4, KH_2PO_4, Glycine

Methods and Procedure

1. Calculation: Calculate the amount of each chemical or stock solution you need to make 10X TAE and 10X TGS buffer and fill in the numbers in the corresponding space in Table 2-1.
2. Make a record entry: In your notebook, you should record the actual mass of the reagent you weighed out on the analytical balance and any actual volumes of reagents added.
3. 10X TAE: Weigh the appropriate amount of Tris-base by the analytical balance and put the powder in a beaker. Then add ~ 50 mL of ddH_2O and the calculated amount of EDTA stock. Go to the chemical hood and use protection to add calculated acetic acid into the beaker. After all the chemicals are dissolved, either by gently shaking your stock flask or using a magnetic stir bar/stir plate combination, use a precalibrated pH meter to adjust the pH to 8.0 by acetic acid or NaOH; then adjust the final volume to 100 mL in a volumetric flask.
4. 1X PBS: Weigh all the components listed in Table 2-1 for PBS buffer and put them in a beaker. Then add ~80 mL of ddH_2O to dissolve the powder. Once it is dissolved, adjust the pH to 7.2 by phosphoric acid or NaOH; then bring the final volume to 100 mL in a volumetric flask.
5. 10X TGS: Add the appropriate amount of Tris-base and Glycine to a beaker; then add the calculated amount of SDS from stock and ~60 mL of ddH_2O to dissolve the powder. Finally, adjust the pH to 8.3 and the volume to 100 mL.

RESULTS AND DISCUSSION

Briefly and succinctly summarize your results. In writing up your results and discussion for this laboratory experiment and others, consider carefully the following: Could you organize your data into a table or make a data figure/plot of it? Are there any discrepancies or inconsistencies or trends in your experiment or data, and if so, can you describe some reasons why? Did you accomplish what you set out to do, or did your question of goals or the way you think about them or your experiment change upon completing the experiment and analyzing your data? What could you have done differently for this experiment to work better?

PROBLEM SET

1. Calculate the number of moles you dissolved of every individual salt and component in each buffer (TAE and PBS).
2. Calculate the molarity of each of these buffer components in units of molar (M).
3. Convert these into mass/volume (wt%).
4. How many types of ions exist in your PBS buffer? What are they?
5. Using the Henderson–Hasselbach equation for Tris and for phosphate and using the final pH and known values for pKa, calculate the [A-]/[HA], which refers to the dissociation into ions.
6. If you mistakenly used 0.5M EDTA instead of Tris, what ramifications would this have for the buffer?
7. If you mistakenly forgot to include EDTA in your TAE buffer, what would be the impact of this on your buffer and its ability to protect biological molecules?

Additional Materials

[1] Rodney Boyer, Biochemistry Laboratory—Modern Theory and Techniques, 2nd ed., Pearson Prentice Hall, 2012, 2006, Chapter 3, General Laboratory Procedures, Section A pH, Buffers, and Biosensors.
[2] Jeffrey Paradis, Kristen Spotz, Hands On Chemistry, McGraw-Hill, 2006, pp. 5-2–5-3.
[3] Terence G. Cooper, The Tools of Biochemistry, John Wiley and Sons, 1987, Chapter 1, Potentiometric Techniques.
[4] John F. Robyt, Bernard J. White, Biochemical Techniques: Theory and Practice, Brooks-Cole Publishing Company, 1987.

Notes

3

Investigating the Physico-Chemical Properties of Amino Acids and Their Analysis by Thin Layer Chromatography

SAFETY AND HAZARDS

The solvents used in this experiment, such as methanol (CH_3OH) and acetonitrile (CH_3CN), which is a chemical derivative of cyanide, are toxic and should be handled with great care in the chemical hood. Care also must be taken to avoid skin contact, inhalation, or ingestion. Wear gloves and work in the chemical hood with these materials, and once they are aliquoted into a 96-well plate, take care to avoid spilling them onto the bench. Ninhydrin, the amino acid stain, is also hazardous and should be used in the chemical hood. You should use gloves while performing this procedure to avoid staining your skin. You should always consult the MSDS sheet for each chemical and reagent used in an experiment, prior to conducting it, in order to keep yourself, your partners, and your lab mates safe when working in the laboratory.

INTRODUCTION

Amino acids are the building blocks of all the proteins in the body. Proteins provide the structure and, in the case of enzymes, the capacity to catalyze all the biochemical and metabolic reactions necessary for life. Each protein has a unique sequence of amino acids, which determines the structure and function of the protein. Although there are 20 kinds of standard amino acids found in eukaryotes, they all share a common basic chemical structure, as shown in Figure 3-1.

FIGURE 3-1   Basic amino acid structure

As shown in the figure, each amino acid exhibits this basic structure, with amine (NH$_3^+$) and carboxylic acid (COO$^-$) groups attached to a central (or alpha) carbon. The structure shows the "zwitterionic" form present at neutral pH, whereby the amino group has a positive charge (+) and the carboxylic group has a negative charge (–). Each amino acid has a unique R chain, according to which the 20 amino acids can be segregated into three groups, as shown in Table 3-1.

As shown in the table, the R chain and its functional groups make each amino acid a unique biochemical. This determines whether the amino acid imparts a nonpolar, polar, or ionic nature to the protein it is present within. The new field of *proteomics* is the study of not only one protein's structure-function, but rather all the proteins existing in the living organism and the interactions among them, which are all essentially governed by the protein's primary (1°) structure or sequence of amino acids. Therefore, it is the amino acids themselves and their chemical properties that, when present within different regions of a protein, allow interaction with each other to enable protein folding into their unique structures three-dimensionally. Once these proteins are folded, it is the accessibility of the various amino acids in the protein and the functional groups present on the surface that determines points of contact and interactions with other biomolecules. Some examples include the formation of a hydrophobic domain that can be inserted into the membrane, an active site for interaction with its specific substrate in the case of enzymes, a leucine zipper or tract of arginine and lysine for interaction with DNA or RNA, and many others. In the end, the amino acid sequence and the interactions it allows are what determine the beautiful complexity of all the proteins in the proteome.

As summarized in Table 3-1, amino acids are categorized into three main classes according to the nature of these R chain functional groups: (1) nonpolar, allowing proteins to create hydrophobic domains or pockets; (2) polar, such as hydroxyl (-OH) or amide (-CONH2), promoting hydrogen bond formation with water, other amino acids, or biomolecules; and finally (3) ionic, either positive (cationic) or negative (anionic), used to form highly stable ionic interactions between amino acids present within the protein, to other biomolecules, substrates, etc.

Among these 20 amino acids, proline (Pro, P) is a special case; since its amino group is part of a ring five-member structure, it is actually an "imino" acid, not an amino acid as all others are. Another special amino acid is cysteine, which is the only common amino acid containing a sulfur or thiol (-SH) in its R chain. The molecular structures of Pro and Cys are shown in Figure 3-2.

TABLE 3-1 Side-Chain Functional Group (R) Categories of Amino Acids

Non-polar	Polar	Ionic
Glycine (Gly, G) R = H	Serine (Ser, S) R = CH_2OH	Lysine (Lys, K) R = $CH_2CH_2CH_2CH_2\text{-}NH_3^+$
Alanine (Ala, A) R = CH_3	Cysteine (Cys, C) R = CH_2SH	Arginine (Arg, R) R = $\text{-}(CH_2)_3NH\text{-}\overset{\parallel NH}{C}\text{-}NH_2$
Phenylalanine (Phe, F) R = CH_2	Tyrosine (Tyr, Y) R = CH_2	Histidine (His, H) R =
Methionine (Met, M) R = $\text{-}CH_2CH_2\text{-}S\text{-}CH_3$	Threonine (Thr, T) R = $\text{-}\underset{OH}{CH}\text{-}CH_3$	Asparate (Asp, D) R = $\text{-}CH_2COOH$
Leucine (Leu, L) R = $\text{-}CH_2CH(CH_3)_2$	Asparagine (Asn, N) R = $\text{-}CH_2\text{-}\overset{\parallel O}{C}\text{-}NH_2$	Glutamate (Glu, E) R = $\text{-}CH_2CH_2COOH$
Valine (Val, M) R = $\text{-}CH(CH_3)_2$	Glutamine (Gln, Q) R = $\text{-}CH_2CH_2\text{-}\overset{\parallel O}{C}\text{-}NH_2$	

Continued

TABLE 3-1 Experimental Side-Chain Functional Group (R) Categories of Amino Acids—cont'd

Non-polar	Polar	Ionic
Isoleucine (Ile, I)		
$R = -CH-CH_2CH_3$		
$\qquad\vert$		
$\qquad CH_3$		
Proline (Pro, P)		
$R =$		
Tryptophan (Trp, W)		
$R = $		

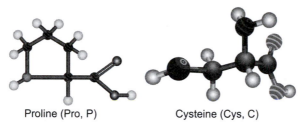

Proline (Pro, P)   Cysteine (Cys, C)

FIGURE 3-2 Molecular structure of proline (obtained from 4vector.com) and cysteine (sciencephoto.com). The blue or green atom represents the nitrogen (N) in the amino group (or imino in the case of Pro). The red atoms represent oxygen (O) within the carboxyl group of all amino acids. The yellow atom represents sulfur, uniquely present in methionine or cysteine, as shown here

For example, cysteine (Cys) allows proteins to form internal or external disulfide bonds (Cys-S-S-Cys), creating covalent linkages within a folded protein or between two chains or subunits in the case of quaternary structure, greatly stabilizing protein structure. The other two amino acids to be used in this experiment are leucine (Leu), an excellent example of an amino acid from the nonpolar class, and lysine, a typical example of an amino acid from the ionic class. Remember, with the exception of Pro, all amino acids have a free amino group, and all have a free carboxyl group. Thus, it is the specific R group (also called the side chain) that distinguishes the 20 different amino acids from one another. To illustrate these differing fundamental characteristics, in this lab you test four amino acids with various side chains and determine their solubility and thin layer chromatography (TLC) R_f and pattern.

TECHNICAL REVIEW

Chromatography is the general practice used to characterize, separate, and purify molecules. Modern chromatography has become automated and robotic over the years, encompassing many different classes of chromatography, including gas chromatography (GC), high-pressure liquid chromatography (HPLC), and fast-pressure liquid chromatography (FLPC). These systems are coupled with various types of detectors. Some examples are LC-MS, which refers to liquid chromatography (LC) with detection by mass spectrometry (MS); UV/Vis, which takes advantage of the ultraviolet (UV) or visible (Vis) light absorbance of many biomolecules; and fluorescence-based detectors, which measure fluorescent signals from reactions or fluorescent molecules attached or conjugated to various biomolecules, and so on. In some cases such as carbohydrates or lipids, which may or may not have a chromophore (a light-absorbing molecular structure), other modes of detection including electrochemical, evaporative light scatter, and others can be used in these special cases.

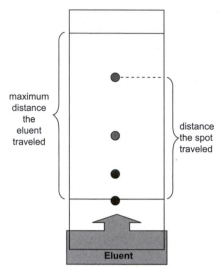

FIGURE 3-3   Sample of marked solvent front and migrated spots

All the aforementioned techniques use some sort of chemically func-tionalized solid support or beads (mm-μm in diameter) packed into a column of varying length and inner diameter to physically separate mol-ecules according to their physical and chemical properties. The separa-tion is based on the strength of the molecules' interaction with the beads either by ionic, hydrophobic, or specific binding forces. When separation is based on hydrophobic interactions, hydrophobic interaction chroma-tography (HIC) is generally used. HIC works by attracting the molecules to be separated to the beads by hydrophobic forces. These molecules can then be mobilized up the plate by some combination of a nonpolar and polar solvent. More polar, ionized molecules do not interact with the plate and travel up the plate very rapidly, whereas more nonpolar molecules are retained and travel fairly slowly, depending on the elution buffer and its polarity. Hence, a characteristic of a biomolecule is a constant or factor related to how far it travels up a given type of TLC plate for a given sol-vent mixture. This is shown schematically in Figure 3-3.

The R_f is calculated as shown in Equation 3-1:

$$R_f = \text{Distance of Solute/Distance of Solvent} \qquad \text{Eq. 3-1}$$

EQUIPMENT HIGHLIGHT

In this experiment, you do not use a column , but instead you coat the silica beads onto a thin layer plate—hence the name thin layer chroma-tography (TLC). TLC is perhaps the simplest form of chromatography,

FIGURE 3-4   Example of TLC chamber setup

especially when coupled with the HIC principle. In a TLC plate, the beads (usually some sort of derivatized silica with various alkyl carbon chains, e.g., C_{18}) are coated onto the surface of the plate or strip.

A typical TLC setup is shown in Figure 3-4. The molecules to be analyzed or separated are spotted near the bottom of the plate, and then the bottom of the plate is immersed in a solvent or solvent mixture (eluent or elution buffer) in a beaker. The distance the molecules travel up the plate is primarily a function of their chemistry, the type of coating on the beads coating the plate, and the type of solvent or elution buffer used.

The experimental setup, materials, and suggested protocol for this experiment are described in the next section.

EXPERIMENT 3

Equipment

- Micropipettes (P1000, P200, P20)
- Drying oven (optional) or blow dryer

Materials

- Amino acids (Leu, Pro, Lys, Cys, plus an unknown amino acid mixture)
- 500 mL beaker
- Parafilm
- Premade ninhydrin staining solution (0.2% ninhydrin in ethanol, e.g., Sigma Ninhydrin spray reagent, Cat# N1286)

- Silica TLC plates (e.g., Millipore classical TLC plates, Cat# 105748) cut into size 6 × 8 cm
- ddH$_2$O
- Methanol (MeOH)
- Acetonitrile (ACN)
- Isopropanol (Ipr-OH)
- Ethanol (ETOH)
- 1M acetic acid (CH$_3$COOH)
- 96-well plates, clear bottom
- Capillaries (optional, e.g., Fisher Cat# 22-260-943)
- Weighing spatula or forceps

METHODS AND PROCEDURE

Solubilizing the amino acids: In this experiment, you utilize a few of the common solvents in use in biochemistry and molecular biology today, including double-distilled and deionized water (ddH$_2$O), methanol (MeOH), ethanol (ETOH), isopropanol (Ipr-OH), and acetonitrile (ACN), or various combinations thereof. These solvents have varying ranges of polarity or nonpolarity; hence, this property affects their ability to solubilize or dissolve the different types of amino acids. You may then experiment with these solvents and, if time permits, try different volume:volume (vol/vol) mixtures of each in order to best dissolve the amino acids as follows:

1. The basic procedure is to obtain a 96-well plate and sketch a design matrix in your lab notebook similar to that shown in Table 3-2.
   Use this portion of the experiment to investigate how well the amino acids dissolve in pure liquid (water, methanol, etc.) versus combinations of liquids. Because it is not necessary to use all 96 wells, skip a row in between, when possible, to avoid contamination.
2. Prepare solvents: Once the plate design is laid out, the next step is to aliquot or pipette 200 microliters (200 µL) of each solvent or mixture into each of the wells. An example of a possible design in terms of which columns or wells in the plate have which solvent or combination of solvents is illustrated in Table 3-3.
   Note: Be sure to include a key for what solvent is contained in each column, similar to Table 3-3.
3. Once the solvents have been aliquoted into each well, the next step is to carefully transfer a small amount of each amino acid into each well. One or two crystals or granules are sufficient per well (< 0.5 mg) and can be transferred by the tip of a small spatula. (Be careful to wipe the tip with a KimWipe or other clean cloth after each transfer.) Pipette the solvent

TABLE 3-2 Sample Design Matrix for Solubility Tests

	1	2	3	4	5	6	7	8	9	10	11	12
1 (LEU)												
2												
3(PRO)												
4												
5 (LYS)												
6												
7 (CYS)												
8 (UNKN)												

(UNKN) = unknown amino acids sample

TABLE 3-3 Sample Key for Design Matrix for Solubility Test

COLUMN 1	COLUMN 3	COLUMN 5	COLUMN 7	COLUMN 9	COLUMN (mixture) 11
ddH$_2$O	MeOH	EtOH	iPr-OH	ACN	ddH$_2$O:ACN (50%:50%)

FIGURE 3-5   Spotting the TLC plate

and amino acids up and down to mix and dissolve, or shake the plate on an orbital plate shaker for a few minutes. In each square of the design matrix, record "S" for soluble, "I" for insoluble, and "P" for partially soluble. This may be easier to analyze if the plate is set on a dark surface.

4. Choose one solvent or mixture that best dissolved each amino acid and record your data in your notebook. Then proceed to the next step, laying out and spotting your TLC plate as described next.

Spotting the TLC plate: Draw a line with a pencil, approximately 1.0 cm above the bottom of the TLC plate. Using a fresh tiny pipette tip or capillary tube for each amino acid dissolved with your chosen solvent, place small, single spots (less than 1 µL) along this line. Keep the diameter of the spots within 2–3 mm. Leave enough space (about 5 mm, depending on size of plates and the number of spots) so that the spots do not run together during elution. *Spot two times to ensure enough amino acids are coated onto the plate. It is important to allow each spot to dry completely before respotting.* An example of spotting the TLC plate is shown in Figure 3-5.

A good visual indicator of a dry spot is when the spots are no longer visible; even spots that appear bright white may not be completely dry. Be sure to make a sketch of your TLC plate in your lab notebook, labeling which spot corresponds to which amino acid.

Eluting the TLC plate: Pipette 5–10 mL of the elution buffer (90%:10% Ethanol:Acetic Acid, 1 M) into a 500 mL beaker. Place the dried TLC plate into the beaker, spotted side up, being careful not to immerse the spots or pencil line into the solvent; then seal the top with parafilm. The TLC plate should rest at an angle against the side of the beaker (refer to Figure 3-4). Almost immediately, the solvent should begin traveling up the TLC plate by capillary action. Allow the solvent to run for about 30 minutes or until the solvent is 85%–90% up the plate. Quickly and carefully remove the plate from the chamber and mark the line where the solvent stopped. This

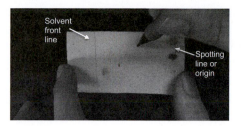

FIGURE 3-6   Marking the spots after ninhydrin staining

is the solvent front (refer to Figure 3-6). This is used in the equation to calculate the R_f value (see Eq. 3-1). Important: Because the elution process is somewhat time-consuming, you should proceed to the TLC portion quickly.

Ninhydrin staining: After the elution is complete and the solvent front is marked, allow the TLC plate to air dry. Once the plate is dry, spray it thoroughly with the ninhydrin spray and place it in the drying oven (~60°C) for 5–10 minutes, or until color develops. Ninhydrin reacts with the amino group and forms an adduct [1]. Most of the amino acids produce a purple or brownish-purple color, although sometimes there is color variation when the side chain or structure allows the formation of a unique product. Remove the plate from the drying oven; then observe and record your results. Using a pencil, mark in the center of each spot, as shown in Figure 3-6, and measure the distance traveled to calculate the R_f value.

RESULTS AND OBSERVATIONS

In writing up the results from this experiment, consider the following points in organizing and presenting your data:

- In your lab notebook, record any differences between the amino acids' migration distance and color (keep the TLC plate or a copy in your notebook).
- Calculate the R_f value for each amino acid (show all calculations).
- It may be helpful to summarize your results in a summary table that compares the amino acids' solubility, R_f, and any other observations.
- Based on the solubility and TLC results and how your standards ran, predict what the unknown amino acid is.

DISCUSSION

Data in biomolecular science experiments can be qualitative or quantitative. For example, the analysis of the amino acid solubility in the various solvents is an example of an empirical determination; it is qualitative, that is, whether the amino acid completely dissolved, only partially dissolved, or was insoluble. No value or measurement number was derived. On the other hand, the R_f value is quantitative. Often in biomolecular science, you make comparisons between groups or sets of data, based on either empirical observations or quantitative measurements.

In writing up a brief and concise discussion based on your results for your notebook, consider the following leading questions or points:

- As far as you could tell, did the amino acids' solubility and/or mobility on the TLC plate follow any type of pattern? If so, briefly and accurately describe it. How would you predict any other amino acids' separation on this TLC plate?
- Was there any difference in color after the amino acids were stained? If so, how do you explain this?
- When you are writing up any experiment, it is inevitable that the experiment may have surprises or not go quite as expected. Sometimes these complications can compromise the results. At the very least, almost every time you perform an experiment, there are things that could have been improved. If you had the chance to do the experiment again, what would you do to improve it? Or if you could design another experiment entirely related to this one but new and improved, what would that look like? Part of your discussion should succinctly summarize the important major points of your results and consider how to improve or expand on the experiment.

PROBLEM SET

1. Based on your R_f values and the pattern observed for a representative nonpolar, polar, and ionic amino acid, what would you predict the R_f for another ionic amino acid such as aspartate might be?
2. Was the color for any of the amino acids in your test set different upon staining with ninhydrin? How might this reflect the difference in its chemical reaction?
3. Bioconjugate chemistry is a relatively new branch of science whereby biological macromolecules, sometimes even from different classes, are connected by a covalent bond or linkage. For example, a lipoprotein or a glycolipid in which part of the molecule is a lipid and part protein or is part carbohydrate and part protein can be considered a bioconjugate, and there are many other examples. After a bioconjugation synthesis, typically medicinal chemists use TLC to determine the extent of their reaction where it can be used to separate the starting reactants from the

product (the bioconjugate). For example, consider the reaction of two cysteine amino acids together in the formation of a disulfide bond and the biological compound cystine (Eq. 3-2):

$$\text{Cys-SH} + \text{Cys-SH} \longrightarrow \text{Cys-S-S-Cys} \ (\text{Cystine}) \qquad \text{Eq. 3-2}$$

Draw or sketch what the TLC pattern might look like where the starting material is separated by the disulfide-bond-linked dipeptide.

4. TLC is still useful in the lab but is being replaced by automated chromatographic methods. For example, TLC, FPLC, and HPLC are now automated processes, and computers can control various parameters, such as flow rate, blends of solvent mixtures, etc. Can you think of how the ability to manipulate the parameters of these automated chromatographic methods (sample size, column plate length, flow rate, solvent gradients, etc.) might accentuate or improve an investigator's ability to separate amino acids or other biomolecules?

5. Since proteins are made up of these 4 plus the other 16 amino acids, how might you take advantage of these properties of the amino acids in the chromatography of a larger protein? How about a complex mixture of proteins?

Acknowledgments

We thank Professor B. Breyfogle, Missouri State University Department of Chemistry, and former students K. J. Flores, J. J. Black, J. L. Bucheit, and D. Louiselle for their excellent technical assistance and optimization of this lab.

Additional Materials

[1] R.L. Lundblad, C.M. Noyes, Chemical Reagents for Protein Modification, vol. 2, CRC Press, 1984.
[2] John F. Robyt, Bernard J. White, Biochemical Techniques: Theory and Practice, Brooks-Cole Publishing Company, 1987, Chapter 4, Chromatographic Techniques.
[3] S.O. Ferrell, L.E. Taylor, Experiments in Biochemistry, A Hands-On Approach, Brooks/Cole Laboratory Series, 2005, Chapter 5, Chromatography.
[4] C.L.F. Meyers, D.J. Meyers, Thin-layer chromatography, in: Current Protocols in Nucleic Acid Chemistry, John Wiley and Sons, 2008.
[5] Preparing your own thin layer chromatography plates and then using them. The Instructables Online. http://www.instructables.com/id/Preparing-your-own-thin-layer-chromatography-plate/ (accessed 28.09.14).

Notes

Rapid Purification, Gel Electrophoresis, and Enzyme Activity Assay of the Luciferase Enzyme from Fireflies

SAFETY AND HAZARDS

As always, you should check the MSDS sheets for all chemicals used in each lab experiment. You should be careful to avoid excessive skin contact or ingestion of any potentially harmful chemical. In addition, electrophoresis represents a possible electrical shock hazard. Follow instructions carefully for this step.

INTRODUCTION

The classical approach is to first purify a protein before studying its structure and function. Bacteria such as *E. coli* are often utilized for this purpose. Genes encoding a protein of interest can be easily inserted into the bacteria via a plasmid vector under the control of a specific antibiotic or inducer, allowing the expression of the desired protein. A typical purification scheme is illustrated in Figure 4-1.

As shown schematically in the figure, the general approach involves expression in *E. coli* (sometimes yeast or other microbes or cells are used), lysis, centrifugation, precipitation, column purification, gel electrophoresis, and activity assay. The host cells are grown up to the desired scale, the cells are lysed, and the desired organelle containing the protein (membrane, vacuoles, cytosol) is isolated by centrifugation. The proteins are then extracted, precipitated, and purified by binding to beads and/or passage through column chromatography prior to gel and activity analysis.

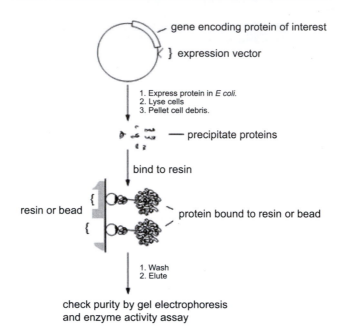

FIGURE 4-1   General approach used for protein purification from *E. coli*

Strains of *E. coli* have been genetically modified for their safety and use in the laboratory for this purpose. These strains grow rapidly on minimal media, and the parameters that affect protein production (temperature, pH, % oxygen, etc.) can be readily optimized. Modern biotechnology companies greatly scale up these processes in huge bioreactors and have large-scale purification equipment in place to purify specific proteins.

TECHNICAL REVIEW AND EQUIPMENT HIGHLIGHTS

In this case, the firefly tail is a specialized organ for producing luminescence (light). Luciferase catalyzes the oxidation of Luciferin substrate. This reaction is shown in Equation 4-1:

$$\text{Luciferin} + O_2 \xrightarrow{\text{Luciferase, ATP, Mg}^{2+}} \text{oxy-Luciferin} + CO_2 + \text{AMP} + \text{diphosphate} + \text{light} \qquad \text{Eq. 4-1}$$

As shown in the preceding reaction, *Luc* protein enzyme utilizes oxygen (O_2) to oxidize its substrate (Luciferin). ATP and magnesium (Mg^{2+}) are required as cofactors in the enzyme's reaction. The product of the reaction is oxy-luciferin, and the by-products are carbon dioxide (CO_2), adenosine monophosphate (AMP), and diphosphate, resulting from the cleavage of the ATP molecule and, most importantly light, measured on the luminom-eter, which essentially is a photon counter.

Cell lysis: Before an enzyme such as Luciferase can be assayed for activity, you must purify it away from proteases and other proteins and biomolecules that could otherwise affect its structure, function, and activity. For an intracellular protein, this is typically done by breaking up the cell wall or plasma membrane with a diluted detergent (SDS, Triton X-100, Brij-35, etc.), followed by some kind of gentle physico-mechanical force (cell scraping, sonication, etc).

Equipment Highlights

Centrifugation: Once the proteins have been "leached out" of the cells, they must be separated from all the other cell components, fragments of the membrane, other organelles, nucleus, ribosomes, etc. This is typically done by centrifuging the crude protein sample, whereby these other unwanted contaminants and parts of organelles are quite dense and heavy relative to dissolved proteins and will sediment to the bottom of the centrifuge tube after centrifugation. Generally, the sedimentation coefficient or Svedberg constant is much smaller for proteins and much larger for organelles, which often contain many densely concentrated proteins and nucleic acids. So a brief (~10-minute) spin (~10,000 rpm) in a microcentrifuge is often all that is needed to get a cytosolic protein away from everything else in the cell or on/in its membrane.

Today the bench-top microcentrifuge has become commonplace in the lab (shown in Figure 4-2). Samples of 0.5 to 1.5 mL can be aliquotted into microcentrifuge tubes or Eppendorfs, balanced and set into the centrifuge opposite each other, and spun at speeds ranging from as little as 500 up to 13,000 or 14,000 revolutions per minute (rpm). In this experiment, since the Luciferase is present in great quantity in the firefly tail after cell lysis and sonication, cell and tissue fragments and organellar contaminants are easily removed by centrifugation. A typical table-top or bench-top centrifuge used commonly in the biomolecular science lab is shown in Figure 4-2.

It is important to make sure that you balance the centrifuge by loading samples containing exactly the same volume of sample in exactly opposite positions within the centrifuge. Otherwise, if samples are not balanced, the centrifuge can become very dangerous when high-speed spins become off-balance.

Precipitation: After spinning in the centrifuge, the released proteins in the supernatant should be transferred to a clean and sterile vial or Eppendorf. At this stage, the supernatant contains mostly small molecules, carbohydrates, RNA, and proteins. The proteins are next concentrated by precipitation. Common protein-precipitating agents include trichloroacetic acid (TCA), ammonium sulfate, or acetone. They selectively precipitate out mostly proteins and leave the majority of the other unwanted biomolecules in

FIGURE 4-2   Bench top centrifuge

solution. Here, you use acetone because its action is rapid and requires no clean-up steps (dialysis, pH adjustment). However, of these three common precipitation methods, acetone (CH_3COCH_3), being an organic solvent, probably causes most proteins to unfold or denature. The consequence for a protein enzyme would be the loss of its activity or ability to catalyze its reaction.

After precipitation, the sample is again centrifuged, a step that pellets or sediments the desired protein fraction in the bottom of the centrifuge tube and leaves undesirable lipids, membrane fragments, and other partially organic soluble small molecules dissolved in the supernatant. At this point, the sample will mainly contain protein and peptides. Contaminants at this stage will mostly include smaller nucleic acids, such as tRNAs and mRNAs. If necessary, these contaminants can be removed by phenol/chloroform; however, because of time constraints in this lab, you do not perform this step. The supernatant is then discarded and the protein pellet redissolved in an appropriate buffer and kept on ice until further use.

Technical Review

Bead technology and column chromatography: At this point most protein purification schemes usually involve some form of chromatography. The proteins are loaded onto a column packed with microbeads, and the proteins elute through the column at different rates depending on the physico-chemical properties of the column and the proteins' interaction with them. Size exclusion chromatography (SEC) or ion exchange (IEX) chromatography are two of the most common types. In the case of SEC, the pores or spaces between the particles determine which proteins will come through or be slowed down trying to get through the matrix or sieve. Alternatively, in IEX, the beads are coated with a charged molecule and the proteins, depending on their charge, either get attracted or repulsed by the beads. Washing the column through with a salt buffer containing ions then allows the elution of the proteins off the column. However, these techniques take too much time to be able to complete in a typical 2–3-hour laboratory session period for this lab. Instead, we previously found zinc oxide nanoparticles can bind and stabilize Luciferase [1]. Here, you use these particles to replace the beads and column step yet quickly concentrate the enzyme so that you can perform gel electrophoresis and bioassay on it as described next.

Gel electrophoresis: Polyacrylamide gel electrophoresis (PAGE) is typically used for most proteins; t he setup and equipment for this technique are shown in Figure 4-3.

Today many "rigs" for this technique are available, and precast gels can be purchased for convenience. The gels are loaded vertically in the apparatus, and pH-balanced running-buffer-containing ions for electrical current and gradient separation are filled into the upper and lower reservoir. The polyacrylamide forms a meshwork sieve similar to column chromatography beads where the proteins essentially separate according to size when sodium dodecyl sulfate (SDS) is present. Small proteins run down to the bottom of the gel, and larger proteins stay near the top. After the electrophoresis is complete, the power supply is turned off, and the connector cords are disconnected, the gel is removed from its holder and stained in Coomassie blue or any of the other protein gel stains in common use and commercially available today. As shown in Figure 4-3, a set of protein molecular weight standards (ladder) is often loaded for comparison. You can then use the distance that each protein band at a known molecular weight migrated on the gel relative to the band corresponding to the Luciferase isolated from the tail of the firefly to estimate the correct Mw of *Luc*. The absence of a band in the lane where the head sample was loaded is one means of confirmation. If time were available, the traditional means to confirm a protein today is to send it off to be sequenced by protein-based liquid chromatography–mass spectrometry (LC–MS). The purified protein can then be assayed for Luciferase activity using a standard commercially available kit and the light generated quantified by luminometry (luminescence spectroscopy).

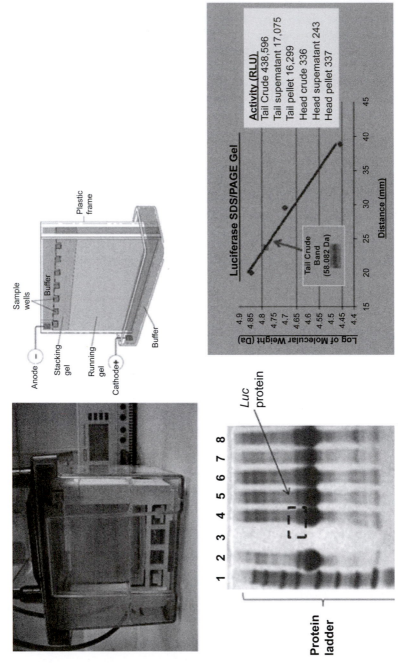

FIGURE 4-3 Gel electrophoresis apparatus, stained gel, and fraction activity analysis

EXPERIMENT 4

Materials and Equipment

- Ice cold acetone (reagent grade)
- Bench-top centrifuge
- Microcentrifuge tubes, plastic mortar, and pestle
- Firefly tail and head to be used as control (1 per bench group, ½ tail or head per group)
- Luciferin (substrate for activity assay)
- Pipetteman and tips
- Sonicator
- PBS buffer substituted with 1% SDS (lysis buffer), PBS (suspension buffer)
- Zinc oxide nanoparticles (Sigma-Aldrich, Cat# 544906-10G)
- Gel electrophoresis supplies (prepoured gradient SDS/PAGE gel; 10X running buffer; loading buffer and power supply; prestained molecular weight ladder; Coomassie blue gel stain)
- Microtiter plate reader or simple luminometer (BMG or molecular devices)
- Enzyme buffer-luciferase reagent (Promega)
- Gel documentation station
- Protein staining buffer—GelCode Blue Thermoscientific Prod# 24592

Methods and Procedure

1. Cut the tail and head into halves and split them with your lab partner. Place them into a 1.5 mL Eppendorf tube; then grind up the tissue with a plastic mortar tip or pipette tip. Pipette in about 0.25 mL PBS/0.05%PBS buffer to each, grind the tissue some more, and then sonicate it for 1–2 minutes.
2. Centrifuge the tubes down in a microcentrifuge (make sure to balance them in the centrifuge beforehand), spinning them down at about 1,000–2,000 rpm for several minutes to pellet the tissue pieces and leave the supernatant containing the various soluble and cytosolic proteins. Decant the supernatant containing the crude Luciferase (from the tail sample) into a new set of tubes (1 for head and 1 for tail).
3. Pipette into each tube about 1 mL of cold acetone, which should fill the 1.5 mL Eppendorfs nearly to the top. You should see the solution turn cloudy if you use a clear Eppendorf tube (the cloudiness means that proteins are coming out of solution in the presence of the organic compounds). Place the tubes into the centrifuge again and

spin down for 1–2 minutes at 10,000–11,000 rpm. After the spin, you should see a pellet at the bottom of your tube (this is the protein). Carefully decant off the supernatant into a new tube and save the pellet, leave the cap open for 5 min to completely evaporate the acetone (residue of acetone will decrease the protein solubility). To the pellet, add about 0.1 mL PBS buffer and gently vortex-mix the sample to dissolve the proteins.

4. Weigh out about 2–3 mg zinc oxide microparticles into a small (0.5 mL) Eppendorf and transfer about 50 μL of your final head and tail solution from step 3 onto the particles. Then vortex-mix for 30 seconds and let sit on ice 1–2 minutes. Spin the particles containing protein down in the microcentrifuge at 10,000–11,000 rpm to pellet the particles and carefully remove the supernatant. Wash the particles in about 20–30 μL of PBS. Keep both fractions for gel and activity analyses.

5. Electrophorese your protein samples. To about 20 μL of head or tail purified sample (after step 4 and/or step 3, depending on how many lanes are available), add an equal volume of protein load buffer. Gently load these samples into each well of the protein gel, leaving room for a protein standard ladder. Fill the top and bottom reservoir of the electrophoresis apparatus with protein running buffer (be sure there is a good seal and there are no leaks). Hook up the power supply to the cathode and anode connections, set for 100–110V, and electrophorese your sample for 45 minutes to 1 hour or however much time remains in the lab session. Using a precolored or prestained set of molecular-weight ladder standard proteins is the best way to judge how long to run the electrophoresis as these bands separate down the gel.

6. While the gel is running, measure the activity of your sample. Remove 10 μL aliquots of your head and tail samples in triplicate and pipette them into three individual wells of a white, opaque, 96-well plate. Make sure to include a set of control wells that contain Luciferase buffer and substrate (but no head or tail sample). Follow the directions for the *Luc* assay kit, adding enough buffer and reagent containing substrate to each well accordingly. It is helpful to take a quick flash on your plate because Luc activity is notoriously short-lived, and the light production half-life is only 5–10 minutes [3,4].

RESULTS AND DISCUSSION

Record your observations and results in your notebook. Consider the following problem set in your results and discussion of this lab.

PROBLEM SET

1. Is there evidence of having successfully purified Luciferase?
2. How do you argue or prove that? How do you demonstrate that?
3. If you see a band in the tail sample on the stained gel, is it at the correct molecular weight based on what is in the literature for Luciferase? Is there such a band in the head sample?
4. Luciferase activity are normally stated in relative light units (RLU), the reading from the luminometer. How many units of activity (RLU) are present in the whole tail sample? The whole head?
5. If the whole tail weighs 0.5 mg, calculate the specific luciferase activity (RLU/mg protein). What assumption did you make here?
6. Now that you have a purified sample of Luciferase, what other kinds of studies or characterization could you perform on it if time were available?

References

[1] S. Barber, M. Abdelhakiem, K. Ghosh, L. Mitchell, R. Spidle, B. Jacobs, et al. Effects of nanomaterials on luciferase with significant protection and increased enzyme activity observed for zinc oxide nanomaterials, J Nanosci Nanotechnol 11 (12) (2011) 10309–10319.
[2] S.O. Ferrell, L.E. Taylor, Experiments in Biochemistry, A Hands-On Approach, Second ed., Brooks/Cole Laboratory Series, 2006, Chapter 4, Enzyme Purification.
[3] Firefly luciferase. http://www.ebi.ac.uk/interpro/potm/2006_6/Page2.htm (accessed 29.09.14).
[4] L.F. Greer 3rd, A.A. Szalay, Imaging of light emission from the expression of luciferases in living cells and organisms: a review, Luminescence 17 (1) (2002) 43–74.

Additional Materials

[1] William Horspool, Francesco Lensi (Eds.), CRC Handbook of Organic Photochemistry and Photobiology, CRC Press, 2004.
[2] John F. Robyt, Bernard J. White, Biochemical Techniques: Theory and Practice, Brooks-Cole Publishing Company, 1987. Chapter 8, Biological Preparations.
[3] Rodney Boyer, Biochemistry Laboratory—Modern Theory and Techniques, Second ed., Pearson Prentice Hall, 2012, 2006. Chapter 6, Characterization of Proteins and Nucleic Acids by Electrophoresis.
[4] O. Gaal, G.A. Medgyesi, L. Vereczkey, Electrophoresis in the Separation of Biological Macromolecules, John Wiley and Sons, 1980.
[5] B.D. Hames, D. Rickwood, Gel Electrophoresis of Proteins, IRL Press, Oxford, UK, 1981.

Notes

Hexokinase and G6PDH Catalyzed Reactions of Glucose Measurement

SAFETY AND HAZARDS

In this lab, you use the micropipette extensively to aliquot samples and reagents. Pipette tips are sharp objects; you should take extra caution during pipetting. Avoid skin contact, inhalation, or in any way ingesting any chemicals or reagents used in the lab.

INTRODUCTION

Carbohydrates are one kind of important biomolecules, besides proteins. Unlike proteins or amino acids, which contain nitrogen, carbohydrates are composed of only carbon, hydrogen, and oxygen, usually with an empirical formula, $C_m(H_2O)_n$. Sugar is a generalized name for a type of carbohydrate, which is usually a sweet-flavored substance. Sugars can be categorized into groups according to the number of sugar rings contained, such as monosaccharides, disaccharide, oligosaccharides, and polysaccharides. Most common sugars are listed in Table 5-1 and Table 5-2.

Glucose is one of the most important carbohydrates in biology. It is one of the three dietary monosaccharides, along with fructose and galactose. Glucose is absorbed directly into the bloodstream during digestion, which is the primary source of energy for the body's cells. The glucose levels in the blood are regulated by negative feedback in order to keep the body in homeostasis. When the blood glucose level is too low, the pancreas releases a hormone called glucagon, which converts glycogen into glucose from the liver. When the blood glucose level is too high, insulin is released from the pancreas. This hormone is able to cause the liver to convert

TABLE 5-1 Common Dietary Monosacchrides

Monosaccharides	Open chain structure	Cyclic isoforms			
D-Glucose		α-D-Glucopyranose	β-D-Glucopyranose	α-D-Glucofuranose	β-D-Glucofuranose
D-Fructose		α-D-Fructopyranose	β-D-Fructopyranose	α-D-Fructofuranose	β-D-Fructofuranose
D-Galacotose		α-D-Galactopyranose	β-D-Galactopyranose	α-D-Galactofuranose	β-D-Galactofuranose

TABLE 5-2 Common Dietary Disaccharides and Polysacchrides

Disaccharides/polysacchrides	Structural components	Chemical linkage between components	Chemical structure
Sucrose	Glucose Fructose	O-α-D-glucopyranosyl-(1→2)-β-D-fructofuranoside	
Lactose	Galactose Glucose	β-D-galactopyranosyl-(1→4)-D-glucose *	
Amylose	Glucose	α(1→4)	

Continued

TABLE 5-2 Common Dietary Disacchrides and Polysacchrides—cont'd

Disaccharides/ polysaccharides	Structural components	Chemical linkage between components	Chemical structure
Glycogen	Glucose	Linear: α(1→4) Branch: α(1→6)	
Cellulose	Glucose	β(1→4)	

*Glucose can be in either α-pyranose form or the β-pyranose form

excessive glucose into glycogen. Additionally, insulin is able to stimulate cells to take up glucose from the blood via the GLUT4 transporter, thus to decrease the blood glucose level. Failure to maintain blood glucose in the normal range leads to conditions of high or low blood sugar. Diabetes is the most common disease related to abnormalities in blood sugar control. Because of this unique health condition indicating functions of blood glucose level, it is always an interesting topic for scientists to investigate how to measure the glucose level accurately and conveniently.

Two major methods have been used to measure glucose. The first one is based on the nonspecific reducing property of glucose to react with a substance that changes color (see Table 5-3, for example). This method may produce erroneous readings in some situations when other reducing agents are present in the sample. More recently, glucose-specific enzymes, such as glucose oxidase and hexokinase, have been used to more accurately determine the glucose concentration [1,2].

Hexokinase, which is the first metabolic enzyme encountered by glucose in the cell, phosphorylates glucose and converts it into glucose-6-phosphate (G6P), which lies at the start of two major metabolic pathways in the cell: glycolysis and pentose phosphate pathway. G6P can also be converted to glycogen or starch for energy storage. In glycolysis, G6P is converted into fructose-6-phosphate by an isomerase, whereas in pentose phosphate pathway, G6P is dehydrogenized into 6-phosphogluconolactone by the glucose-6-phosphate dehydrogenase (G6PDH), and NADPH is produced in physiological conditions ($NADP^+$ can be substituted by NAD^+ in an *in vitro* reaction). In this lab, you apply the hexokinase and G6PDH catalyzed enzymatic reaction of glucose to measure the glucose concentration in the samples. However, the glucose level cannot be measured directly; instead, you measure the consequent increase of NADH by reading the absorbance at 340 nm, which reflects the concentration of glucose in the samples (see Table 5-4).

TECHNICAL REVIEW

Dilutions

In this lab, you prepare glucose standards and sample dilutions. To prepare a diluted working solution with a specific concentration, you use the equation $C1 \times V1 = C2 \times V2$ to calculate how much stock solution to add.

TABLE 5-3 Alkaline Copper Reduction Method

Step 1	Glucose + cupric → Cuprous + gluconic acid
Step 2	Cuprous + Phosphomolybdic acid → cupric + Phosphomolybdenum oxide (blue end product)

TABLE 5-4   Hexokinase and G6PDH Catalyzed Reactions of Glucose Measurement

Step 1	Glucose + ATP $\xrightarrow{\text{Hexokinase}}$ Glucose-6-phosphate + ADP
Step 2	Glucose-6-phosphate + NAD $\xrightarrow{\text{G6PDH}}$ 6-phosphogluconate + NADH

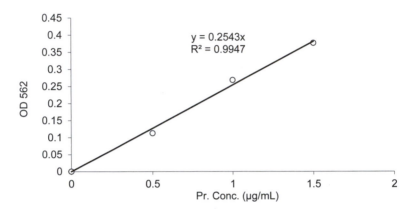

FIGURE 5-1   Standard curve example showing protein concentration determination

Then you add the remaining volume with solvent. To do a 10X dilution, for example, you add 1 volume of stock solution into 9 volumes of solvent. In this case, the stock solution is either 1 mg/mL glucose solution, or the samples (cell culture media or juice), and the solvent is water.

Standard Curves

Standard curves represent the relationship between two quantities. They are used to determine the value of an unknown quantity (glucose concentration) from one that is more easily measured (NADH level). An example of a standard curve for protein concentration determination is illustrated in Figure 5-1.

To calculate the sample concentration based on the standard curve, first you find the concentration for each sample absorbance on the standard curve; then you multiply the concentration by the dilution factor for each sample. See the data-handling process in the example of protein concentration determination in Table 5-5 and Figure 5-1.

UV/Vis Spectrophotometry

In general, spectroscopy refers to a collection of instrumental techniques wherein molecules absorb electromagnetic radiation of a specific

TABLE 5-5 Data-Handling Process of Protein Concentration Determination

ORIGINAL DATA

BSA Conc. (mg/mL)	0	0.5	1	1.5
Measure 1	0.337	0.481	0.674	0.786
Measure 2	0.388	0.473	0.63	0.718
Measure 3	0.395	0.504	0.622	0.746
Calculated mean	0.373333333	0.486	0.642	0.75
Samples	5X Smpl A	10X Smpl A	10X Smpl B	2X Smpl B
Measure 1	0.644	0.506	0.734	0.987
Measure 2	0.649	0.504	0.705	0.956
Measure 3	0.619	0.49	0.71	0.934
Calculated mean	0.637333333	0.5	0.716333333	0.959

BACKGROUND SUBTRACTED (BACKGROUND ABS = 0.3733)

BSA Conc. (mg/mL)	0	0.5	1	1.5
Measure 1	−0.0363333	0.1076667	0.3006667	0.4126667
Measure 2	0.0146667	0.0996667	0.2566667	0.3446667
Measure 3	0.0216667	0.1306667	0.2486667	0.3726667
Calculated mean	0	0.11266667	0.26866667	0.37666667
Samples	5X Smpl A	10X Smpl A	10X Smpl B	2X Smpl B
Measure 1	0.2706667	0.1326667	0.3606667	0.6136667
Measure 2	0.2756667	0.1306667	0.3316667	0.5826667
Measure 3	0.2456667	0.1166667	0.3366667	0.5606667
Calculated mean	0.264000033	0.1266667	0.343000033	0.5856667

CALCULATED CONCENTRATION BASED ON THE STANDARD CURVE (mg/mL)

	1.038144056	0.498099489	1.34880076	Unable to be calculated

ORIGINAL CONCENTRATION BEFORE DILUTION (mg/mL)

	5.2	4.98	13.49	N/A

AVERAGED ORIGINAL CONC. (mg/mL)

	Smpl A	5.1 mg/mL		
	Smpl B	13.5 mg/mL		

wavelength. This depends on their unique atomic composition, bonds, or chemical functionality. Molecules then emit a wavelength that reveals unique information about their chemistry and structure. There are many different kinds of spectroscopy (infrared, Raman, atomic absorption, etc.); however, the previously expensive and less accessible *mass spectrometry* and *nuclear magnetic resonance spectroscopy* are becoming increasingly popular and more affordable, shedding "light" on biomolecules and their interactions. Today, fluorescence and ultraviolet/visible spectroscopy are routinely used as tools to study biomolecules. You will gain hands-on experience with and receive good exposure to these tools in today's lab.

Most spectrophotometers function by having an illumination system or a filter/wheel set that can control the wavelength generated. The specific wavelength is then directed by a series of filters or mirrors to a sample compartment in which light is directed in and through the sample; the light emanating from the sample is collected, its signal transformed, and then subsequently amplified by a detector. The absorbance is then calculated as $A = (I_0 - I_t)/I_0$ and the result is then displayed, usually on a computer monitor (see Figure 5-2).

Beer–Lambert Law

For today's measurement, you do not use the Beer–Lambert equation to determine the glucose concentration; instead, you use a standard curve. However, the classic Beer–Lambert law is certainly a very helpful tool in the biomolecular lab and will help you understand today's experiment better:

$$\textbf{Beer–Lambert equation}: A = \varepsilon * c * l$$

This law states that a molecule's *absorbance* (A) is a function of its *concentration* (c, expressed in units of M; moles/liter), the *path length* (l) of the cuvette (constructed of an optically transparent material,

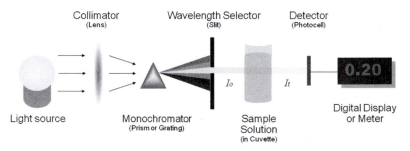

FIGURE 5-2   Structure of a spectraphotometer

usually quartz), and the *molar absorption coefficient* (ε), ε being essentially the biomolecule's efficiency in absorbing light. ε is intrinsic to each biomolecule and is a unique physico-chemical parameter indicating its ability to absorb light at its individual optimal wavelength. The ε of NADH at the wavelength of 340 nm is $6{,}220\ M^{-1}cm^{-1}$. Figure 5-3 shows the absorbance spectra of NAD$^+$ and NADH, where absorbance of 340 nm is well correlated with NADH concentration but not with NAD$^+$ (see Figure 5-4).

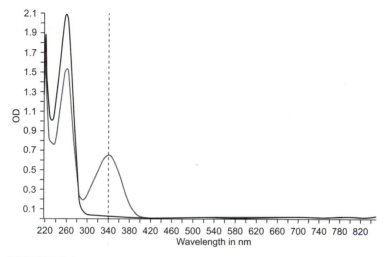

FIGURE 5-3   Absorbance spectra of NAD+ (blue line) and NADH (red line)

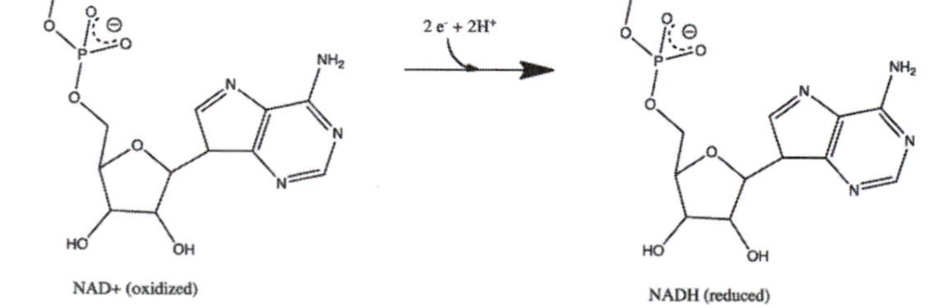

NAD+ (oxidized)   NADH (reduced)

$2\,e^- + 2H^+$

FIGURE 5-4   Reaction from NAD+ to NADH

EQUIPMENT HIGHLIGHT

Microplate Reader

Microplates are flat plates containing multiple wells, which can be used for cell cultures or reaction compartments during analytical research. Most commonly used microplates in basic scientific research are 96-well plates. There are other kinds of microplates, such as 384-well plates and 1536-well plates, which are more widely used in high-throughput screens. Different microplates may also have different levels of transparency in the walls or bottoms: some are clear, some are opaque, some are black, to meet with each measurement condition.

Microplate readers are laboratory instruments designed to detect reaction events of samples. In a 96-well plate, samples are usually in a liquid form, with volume between 100 and 200 µL in each well. Reaction events can be detected by absorbance, fluorescence intensity, luminescence, and so on. In this particular lab, you detect the NADH level generated during the reaction by measuring the absorbance change of 340 nm.

EXPERIMENT 5

Equipment

- Microplate reader suitable for measuring absorbance at 340 nm
- Computer with Microsoft Office

Materials

- 96-well plates, clear bottom
- Glucose assay kit (sigma GAHK-20), which contains glucose standard (1 mg/mL), enzymatic mixture of hexokinase, G6PDH, NAD, and ATP
- Micropipettes and tips
- Samples (soda/juice, cell culture medium)

Methods and Procedure

1. Prepare glucose standards for the standard curve.
   Form a group of four students. Each group prepares one set of glucose standards from the stock (1 mg/mL) using ddH₂O (see Table 5-6).
2. Each pair of students prepares four samples from the given soda/juice and culture media using ddH₂O to dilute (see Table 5-7).

TABLE 5-6　Glucose Standards Preparation

Name	Volume	Concentration
SD0	25 μL	0 μg/μL
SD1	25 μL	0.125 μg/μL
SD2	25 μL	0.25 μg/μL
SD3	25 μL	0.5 μg/μL
SD4	25 μL	0.75 μg/μL
SD5	25 μL	1.0 μg/μL

TABLE 5-7　Sample Preparation

Name	Volume	Dilutions
JU1	100 μL	Dilute 10X
JU2	100 μL	Dilute 50X
CM1	100 μL	Dilute 10X
CM2	100 μL	Dilute 40X

TABLE 5-8　Loading Map of Standards and Samples on a 96-well Plate

SD0	JU1		JU1								
SD1	JU1		JU1								
SD2	JU2		JU2								
SD3	JU2		JU2								
SD4	CM1		CM1								
SD5	CM1		CM1								
	CM2		CM2								
	CM2		CM2								

3. Load standards (10 μL/well) and samples (10 μL/well) to the 96-well plates. For students (two pairs) share one plate (see Table 5-8).
4. Add the enzymatic mixing reagent (100 μL/well) to each well, *except those shaded gray in Table 5-5*. Add 100 μL of ddH$_2$O to each well shaded gray in Table 5-5; these wells serve as the blank absorbance for the sample.
5. Incubate for 15 minutes at room temperature. Then read the absorbance at 340 nm under the guidance of the instructor.

6. Record your data and organize it into a table.
7. Generate a standard curve based on your group's standards concentration and absorbance. (*If your empty wells have absorbance at 340 nm, you should subtract all your absorbance from the empty well absorbance before you do any calculation.*)
8. Calculate your sample concentration based on your standard curve [3]. The total blank must take into account the contribution of the sample and the glucose assay reagent to the absorbance.

$$A_{Sample} = A_{Test} - A_{Sample\ Blank} - A_{Reagent\ Blank}$$

$A_{Sample\ Blank}$: gray wells

$A_{Reagent\ Blank}$: SD0

RESULTS AND DISCUSSION

In your discussion, you might consider the following questions. This experiment resembles which steps of glucose metabolism in the human body? Why do you need to measure the blank of the sample without adding the enzyme mixture? Is there an alternative way to measure the glucose concentration using a similar reagent but without a standard curve? What equipment and factors do you need for your measurement and calculation?

PROBLEM SET

1. What if 1 mole of glucose can generate 2 moles of NADH? How would that affect the calculation, if you use the method of this lab to measure sample's glucose concentration? How would that affect the calculation if you use Beer-Lambert equation to determine the glucose concentration?
2. A student measured the glucose concentration of the cell culture media without any dilution. What type of result would you expect? Do you think the measure is accurate?

References

[1] R.J.L. Bondar, D.C. Mead, Evaluation of glucose-6-phosphate dehydrogenase from Leuconostoc mesenteroides in the hexokinase method for determining glucose in serum, Clin. Chem. 20 (1974) 586–590.
[2] D.A.T. Southgate, Determination of Food Carbohydrates, Applied Science Publishers, London, UK, 1976.
[3] Sigma bulletin product information product code GAHK-20.

Notes

6

Polymerase Chain Reaction (PCR)

SAFETY AND HAZARDS

As always, you should check the MSDS sheets (now available via the Internet) for more information. For example, ethidium bromide (EtBr) and other nucleic acid stains are known carcinogens and should be handled without direct skin contact. Also, all the waste containing EtBr should be disposed according the lab guidelines. Pregnant women should take extra precautions to avoid exposure when using this chemical. Please see the instructor beforehand if you have concerns with this matter. Also, the electrophoresis unit represents a potential electrical shock hazard, and care must be taken to use the instrument in a closed position and to avoid contact with live electrical leads when attached to the power supply. The PCR machine may be hot when it is running or just finished, so be careful when placing or taking tubes out of the reaction tray.

INTRODUCTION

Polymerase chain reaction (PCR), invented by scientist Kary Mullis in the early 1980s, and for which he won a Nobel Prize in 1993, allows researchers to amplify pieces of DNA by several orders of magnitude. This technique has revolutionized many aspects of current research, including DNA cloning and sequencing, functional analysis of genes, the diagnosis of hereditary or infectious diseases, the identification of genetic fingerprints, and so on. The basic components of a PCR reaction include a DNA template, primers, nucleotides, DNA polymerase, and a buffer.

The *DNA template* usually is your sample DNA, which contains the DNA region to be amplified. It could be plasmid DNA, genomic DNA, or even a small amount of tissue. The template DNA is typically given at very low concentrations in a PCR reaction, 1 pg–1 ng of plasmid or viral templates, 1 ng–1 µg of genomic templates.

Primers are short oligonucleotides of DNA (typically 15–25 nucleotides) with a specific sequence that is custom synthesized on an automated DNA synthesizer. Today primers are typically obtained by providing the required sequence to one of many companies that specialize in oligomer provision. Primer design is critical for a successful PCR reaction. In general, the two primers match to the two ends of the segment of DNA you want to amplify. Through complementary base pairing, the 5 end primer matches to the top strand at one end of your segment of interest, and the other primer matches to the bottom strand at the other end, oriented as shown in Figure 6-1. Besides the complementary sequence on the primer, you can also add an extra sequence (such as restriction cutting site, tag sequences, and so on), on the 5 end of the primer, depending on the needs of the experiment.

FIGURE 6-1   Primer design

DNA polymerase is an enzyme complex that amplifies DNA during cell cycle in a living organism. The DNA polymerase used in a PCR reaction usually can tolerate high temperature (95°C), the temperature necessary to separate two complementary strands of DNA in a test tube. For example, the Taq polymerase purified from *Thermus aquaticus*, a strain of bacteria living in a hot spring, can survive near boiling temperatures, and it works quite well at 72°C.

Nucleotides are the building blocks for making the DNA molecules. In PCR reactions, a mixture of four types of nucleotides (ATP, CTP, GTP, TTP; known as dNTPs) will be added. DNA polymerase grabs the complementary nucleotides that are floating in the liquid around it and attaches them to the 3 end of a primer and pairing with the template DNA.

PCR buffers help to maintain the right pH during the reaction cycles and provide necessary ions for enzymes to work. A typical PCR buffer stock solution is provided in a 10X or 5X format; you would need to dilute it to 1X in the PCR reaction. (An example of stock and working solution with the components for a typical PCR reaction is provided in Table 6-1).

TABLE 6-1   Components and Their Concentrations in a Typical PCR Reaction

Components	Stock conc.	Working conc.	Vol. to add
DNA template	10 μg/mL	20 ng/reaction	2.0 μL
Primer 1	20 μM	1 μM	2.5 μL
Primer 2	20 μM	1 μM	2.5 μL
dNTP	10 mM	0.2 mM	1.0 μL
Taq polymerase	1 U/μL	1 U/reaction	1.0 μL
PCR buffer	10X	1X	5.0 μL
water			36.0 μL
reaction vol.			50.0 μL

PCR cycles: A typical PCR procedure takes place in an automated thermal cycler machine consisting of a series of 20–40 repeated cycles with consistent temperature changes. In each cycle, there are three steps called the denaturation step (94°C–98°C), annealing step (50°C–65°C), and elongation step (72°C). On top of the cycling steps, there are usually a single temperature step called the initialization step at a high temperature (> 90°C) before the cycling starts, a final elongation step (70°C–74°C), and a final hold step (4°C) after the cycling ends. Table 6-2 provides guidelines to program the automated thermal cycler.

TABLE 6-2   General Guidelines for Programing a Typical PCR Reaction in an Automated Thermal Cycler Machine [1]

Cycles	Time	Temperature	Steps
1	5 min	95°C	Initialization
~30	30 sec	94°C	Denaturation
	30 sec	55°C–65°C	Annealing
	~ 1 min/kb	72°C	Elongation
1	10 min	72°C	Final elongation
1	∞	4°C	Final hold

TECHNICAL REVIEW

Primer design: Figure 6-2 shows two primers that have been tested to amplify the 14-3-3 gene from *Dictyostelium* cells.

Primer length: It is generally accepted that the optimal length of PCR primers is 18–22 bp. This is long enough for adequate specificity and short enough for primers to bind stably to the template at the annealing temperature.

The *annealing temperature* can be calculated depending on the length and composition of the primers.

GC content <= 50%, 55°C–60°C
GC content >= 50%, 60°C–65°C [1]

MgCl₂: Mg^{2+} binds to the alpha phosphate group of dNTPs and helps in the removal of beta and gamma phosphate from dNTPs. A typical PCR reagent kit provides a 10X buffer and a separate vial of $MgCl_2$, so you can

Dictyostelium's 14-3-3 cDNA sequence:

```
ATGACCAGAGAAGAAAATGTTTATATGGCTAAATTAGCTGAACAAGCTGAAAGATATGAA
GAAATGGTCGAAGCAATGAAAAAAGTTGCTGAATTAGATGTTGAATTAACTGTTGAAGAA
CGTAATCTTTTATCAGTTGCTTATAAAAATGTTATTGGTGCTCGTCGTGCCTCATGGAGA
ATCATCTCATCAATTGAACAAAAGAAGAATCCAAAGGTAATGAAAACCACGTTAAAAAG
ATCAAAGAATACAAATGTAAAGTCGAAAAGGAACTTACCGACATTTGTAATGATATCCTC
GAAGTTTTAGAATCTCACTTAATCGTTTCATCAGCCTCTGGTGAATCAAAAGTTTTCTAC
TACAAAATGAAAGGTGATTACTTCCGTTATTTAGCTGAATTCGCCACCGGTAACCCACGT
AAAACTTCCGCTGAATCATCATTAATTGCCTACAAAGCTGCCTCTGATATCGCTGTCACT
GAATTACCACCAACCCACCCAATCCGTTTAGGTCTTGCATTAAACTTTTCAGTTTTCTAC
TATGAAATCTTAAACTCCCCAGACAGAGCTTGCAACTTAGCCAAGACTGCATTCGATGAT
GCCATTGCCGAATTAGATACCCTCTCTGAAGATTCATACAAAGATTCAACTCTCATTATG
CAATTATTACGTGATAATCTTACTTTATGGACCTCTGACGTTCATAATATGGAAAAAAAT
CAAGATGGTGACGACGATCAAAATGAACCAGGAATGTAA
```

5' primer: 5'AAA AAA GCG GCC GCC ATG ACC AGA GAA GAA AAT G3'

3' Primer: 5'AAA AAA GTC GAC TTA CAT TCC TGG TTC ATT TTG 3'

☐ Sequences complementary to two ends of template DNA

■ Restriction enzyme sequences

▨ Extra hangover sequences for enzymes to grab

■ Added base to prevent codon shift

FIGURE 6-2 Primer design example for amplifying *Dictyostelium's* 14-3-3 cDNA sequence.

optimize the concentration of Mg^{2+} for specific reactions. A standard concentration of $MgCl_2$ in PCR is 2 mM [1].

Enzymes are very sensitive to temperature, so they are stored at $-20°C$ in a non-frost-free freezer, typically in 50% glycerol. You should always wear gloves when handling enzyme tubes. Before opening a tube of enzyme, spin it briefly so that the liquid on the cap and the wall of the tube is collected at the bottom of the tube. Always use a new, clean pipette tip every time you aliquot enzyme.

Equipment Highlight

A thermal cycler machine is the most commonly used machine in the laboratory for PCR reactions (see Figure 6-3). It is also used for some temperature-sensitive reactions. The machine has a thermal block and tube holders. The thermal block is able to increase or decrease the temperature in discrete, preprogrammed steps. Most thermal cycler machines have a touch screen, so you can create, edit, or save a program.

Answer key:

Keys:

The gene you amplified from lambda phage is called capsid component), ID AAA96540.1, your fragment length is 7160-6135+1+12+14=1052 bp

> **CDS:** AAA96540.1
> **Title:** E (capsid component)
> **Location:** 6,135..7,160
> [*Length*]
> **Span:** 1,026

FIGURE 6-3   Thermocycler machines: Bio-rad iCycler

EXPERIMENT 6

Equipment

- Thermal cycler machine
- Micropipettes (P20, P200, P1000)
- Power supply
- Gel apparatus
- Microwave
- Gel documentation system

Materials

- PCR tubes
- 2X PCR master mix
- Primers:
  5 end primer (Sal I site):
  AAAAAAGTCGAC **ATGTCGATGTACACAACCGCCC**
  3 end primer (Not I site):
  AAAAAAGCGGCCGC **TTACGCCAGTTGTACGGACAC**
  (The underlined sequences are the template complementary
  sequences)
- Lambda phage DNA as DNA template
- Nuclease-free water
- Agarose powder
- TAE buffer
- Loading dye
- DNA ladder
- 250 mL flask
- 100 mL cylinder
- Ethidium bromide (1 mg/ml solution premade by your instructor)

Methods and Procedure

1. Prepare a PCR reaction mixture. Calculate how much of each reagent
   you need to prepare one PCR reaction mix (see Table 6-3).
   Add the calculated amount of water into a PCR tube first; then add
   each reagent one by one according to your calculation. After you
   finish adding, use a P20 to gently pipette the mixture up and down to
   mix the reaction.
2. Calculate your primer's annealing temperature according to the
   guidelines presented in this lab, using the underlined sequence
   from both primers. Then design a program for your PCR in the
   thermocycler machine. Next, label your tube clearly, place it in the
   machine, and start the run.

TABLE 6-3   PCR Mixture Preparation

Components	Stock conc.	Working conc.	Vol. to add
DNA template	10 µg/mL	400 ng/reaction	
Primer 1	20 µM	1 µM	
Primer 2	20 µM	1 µM	
2X PCR master mix	2X	1X	
water			
reaction vol.			25 µL

3. While waiting for your reaction, you can prepare an agarose gel (1%). Briefly, weight 0.5 g of agarose powder and dissolve it with 50 mL of TAE buffer in a flask. Heat in the microwave to melt the agarose, and when it is cooled but not yet solidified, add 2 µL of EtBr to the solution and mix well. Then pour the gel mix into the horizontal plastic molds sealed with rubber ends. Allow the gel to cool and harden, making sure you have inserted a loading comb into its upper edge. For more information about DNA/RNA agarose gel, see the equipment highlight in next lab.
4. Determine the size of your target sequence by using the BLAST in NCBI website (http://www.ncbi.nlm.nih.gov/).
5. When PCR reaction is finished, take 5 mL of your reaction, mix with 1 mL of 6x loading dye and load it to the well on the agarose gel. Load 6 mL of DNA ladder to one of the wells. Then, carefully pour the 1X TAE running buffer into the chambers of the electrophoresis unit, just barely covering the top and edges of each side of the gel. Close the lid of the unit before connecting the electrical leads so as to protect yourself from possible electrocution. Run the gel for 30–45 minutes at 100V. Carefully and gently transfer your gel into the gel documentation station. Image it, saving a copy for inclusion in your notebook. (If the students do not have enough time to run the gel, the instructor may help to finish the gel running and document the gel. Students will run an agarose gel again in the next lab.)

RESULTS AND DISCUSSION

How many fragments do you see in your PCR products? What are they? If you do not see your targeting fragment, what would you do differently to make the PCR successful?

PROBLEM SET

Perform the following activities in a computer lab.

1. Go to the NCBI website: http://www.ncbi.nlm.nih.gov/.
2. Under "Popular Resources," click "BLAST."
3. Under "Basic BLAST," click "Nucleotide blast."
4. Copy and paste the following primer sequence into the sequence window:
   ATGTCGATGTACACAACCGCCCTTACGCCAGTTGTACGGACAC
5. In the organism window, type in **Lambda phage (taxid:10710).**
6. Click the "BLAST" button at the bottom.
7. Once the Blast result comes out, look for the hit with 100% match in lambda phage genome.
8. Click on the hit and then determine how big your PCR fragment is.

Reference

[1] Frederick M. Ausubel, Current Protocols in Molecular Biology, vol. 3, John Wiley and Sons, 1995, Chapter 16.

Notes

Investigating Protein:Nucleic Acid Interactions by Electrophoretic Mobility Shift Assay (EMSA)

SAFETY AND HAZARDS

As always, you should check the MSDS sheets (now available via the Internet) for more information about chemicals used in the lab. For example, ethidium bromide (EtBr) and other nucleic acid stains and dyes are known carcinogens. Wear gloves and dispose of these gloves and EtBr properly. Pregnant women should take extra precautions to avoid exposure when using this chemical. Please see the instructor beforehand if you have concerns with this matter. Also, the electrophoresis unit represents a potential electrical shock hazard, and care must be taken to use the instrument in a closed position and to avoid contact with live electrical leads when attached to the power supply.

INTRODUCTION

Protein:nucleic acid interactions play a major role in coordinating cellular functions, and this is especially important in regulating gene expression. For example, during the processes of replication, transcription, and translation, a whole host of proteins and enzymes bind to DNA or to RNA, and are integrally involved in copying the DNA, transcribing it into RNA, or translating the mRNA into protein. As just one example of many in the literature, single-stranded binding (SSB) protein is involved in DNA replication, and the effect of chemically modifying the DNA phosphodiester backbone on interaction with SSB has been well explored by EMSA [1].

More recently, DeLong's group studied the formation of RNA nanoparticles by the binding of protamine to transfer RNA (tRNA), where the complexation of protamine to RNA could also be demonstrated by gel shift [2], as shown in Figure 7-1.

In the gel shown in the figure, the RNA is stained by the commonly used fluorescent dye ethidium bromide (EtBr). When the RNA is bound by the cationic protamine protein, the normally anionic RNA shifts upward toward the negative pole because of the change in its charge/mass.

In this lab, you study the interaction of protamine with plasmid DNA. Plasmid DNAs usually are small, circular, double-stranded DNA molecules, which can be replicated independent of host chromosomes. Here, you use a plasmid DNA (size 6 kb) encoding a green fluorescent protein (GFP) and its interactions with protamine. Protamines are typically isolated from tuna or salmon sperm and are thought to act somewhat like histones in being able to bind and to condense high molecular weight DNA [3]. Protamines are relatively small (5–6 kDa or 5,000–6,000 g/mol) and in comparison to most proteins are highly rich in the positively charged amino acid arginine (Arg, R). Protamine is thus thought to be powerfully attracted to the negatively charged phosphodiester backbone of RNA or DNA, their interaction presumably greatly stabilized through these electrostatic interactions.

The equation for this interaction can be modeled as

$$DNA + \_\ protamine \rightarrow DNA{:}protamine_x \qquad Eq.\ 7\text{-}1$$

where x is the number of protamine molecules bound to DNA. In today's lab, in addition to demonstrating the complex between DNA and protamine, you attempt to determine how many protamine molecules are associated with each molecule of plasmid DNA.

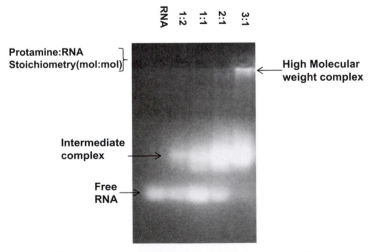

FIGURE 7-1   Protamine:RNA interaction as demonstrated by EMSA

EQUIPMENT HIGHLIGHTS

Gel electrophoresis equipment was shown in Lab 4. Here, you use an agarose gel that is generally used on a horizontal rather than a vertical gel apparatus. You pour your own gel and use a 1% weight/volume (wt/vol) gel, where 1 gram of electrophoresis-quality agar is melted in 0.5–1X TAE buffer (made previously in Lab 2). You must take care that the gel cast is well sealed prior to pouring the molten agarose/TAE solution and to cool it until you feel the bottle is warm prior to pouring (see Figure 7-2).

You then put a "comb" into the gel cast after pouring the liquid agarose solution into it. Once the gel is solidified, you remove the comb so that wells form on the gel to which samples are applied. Then you assemble the DNA and DNA:protamine complexes, add the loading buffer, and carefully load the samples onto the gel prior to electrophoresis, as shown in Figure 7-3.

Finally, you place the gel into the electrophoresis apparatus and electrophorese it for 45 minutes to 1 hour to allow the DNA to separate from the DNA:protamine complexes.

EXPERIMENT 7

Materials and Equipment

- Horizontal gel electrophoresis apparatus and power supply
- Gel documentation station (or UV lamp, protective shield, and camera)
- Electrophoresis-grade low-melting-temperature agarose
- Protamine sulfate, Sigma P3369-10G (2 mg/mL dissolved fresh in water)

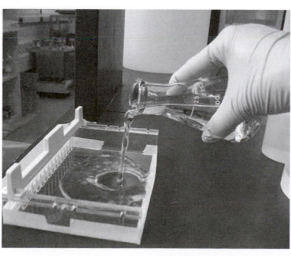

FIGURE 7-2   Image of pouring and agarose gel (Free Online MIT Course Materials)

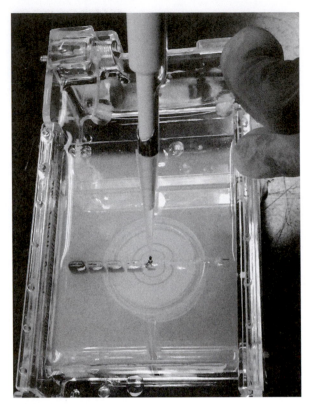

FIGURE 7-3   Loading DNA onto agarose gel for electrophoresis

- Plasmid DNA (for example, Aldevron gWiz GFP Cat# 5006)
- Microcentrifuge Eppendorf tubes
- 1X TAE buffer (40 mM Tris-acetate, 1 mM EDTA, pH 8.3)
- 5X Loading buffer stock (40% sucrose)
- Loading buffer with tracking dye (0.1% Bromophenol blue (BPB) + 0.1% Xylene cyanol)
- Ethidium bromide (1 mg/mL solution premade by your instructor)

Methods and Procedures

1. Prepare a 1% agarose (0.5 g agarose/50 mL TAE buffer) gel. Heat in the microwave to melt the agarose, and when it is cooled but not yet solidified, add 1 μL of EtBr to the solution and mix well. Then pour the gel mix into the horizontal plastic molds sealed with rubber ends. Allow the gel to cool and harden, making sure you have inserted a loading comb into its upper edge. Other options include adding the comb into the middle to allow different separations to occur.

2. In the meantime, prepare a series of samples each containing 2 μL of pDNA. One sample should be a DNA-alone control. The rest of them should have progressively more protamine in them—for example 1, 2, 5, and 10 microliters of the protamine stock.

3. Next, add enough double-distilled, deionized water (ddH$_2$O) so that all samples have a common final volume (such as 15 μL). Then add loading buffer such that the final is 1X (for a 15 μL volume, this would be about 4 μL of a 6X stock).

4. Load the samples into individual wells of the gel (the outside or leftmost well could be just loading buffer with tracking dye). Note here: In our experience, BPB dye interferes with protamine interactions, and although it is more difficult to see, loading buffer without BPB should be used for samples of complexes with protamine. Be careful to change pipette tips after each transfer so as not to contaminate your samples.

5. Carefully pour the 1X TAE running buffer into the chambers of the electrophoresis unit, just barely covering the top and edges of each side of the gel. Close the lid of the unit before connecting the electrical leads so as to protect yourself from possible electrocution. Run the gel for 30–45 minutes at 100V.

6. Carefully and gently transfer your gel into the gel documentation station. Image it, saving a copy for inclusion in your notebook. (Remember to record the order of your samples.)

RESULTS AND DISCUSSION

Record your observations and results in your notebook. Explain situations in which you saw a gel shift or in which you did not. Was there any change in the banding pattern or staining intensity? Given the gel mobility shift pattern, what can you say about the complexes?

PROBLEM SET

1. Find the sequence for protamine and determine how many arginines it has.

2. Based on the predicted sequence for protamine, what is its overall charge expected to be at neutral pH?

3. At the amounts of protamine added in each tube and based on a 2 mg/mL stock, what was the number of moles and the molarity in units of molar (M) and millimolar (mM) of protamine in each tube?

4. Based on the GFP plasmid used in this experiment and assuming an equal ratio of A, T, G, and C and that the plasmid is 5757 base pairs, to achieve charge neutrality, calculate what number of protamine molecules are bound to each DNA molecule.

5. Based on this molar ratio, what is the overall expected charge of the complexes assuming one negative charge is provided per base pair of the plasmid.

6. Can you think of another technique that could be used to probe these complexes that would be more insightful and/or give you additional information about them?

References

[1] X. Cheng, R.K. DeLong, E. Wickstrom, M. Kligshteyn, S.H. Demirdji, M.H. Caruthers, R.L. Juliano, Interactions between single-stranded DNA binding protein and oligonucleotide analogs with different backbone chemistries, J. Mol. Recognit. 10 (2) (1997) 101–107.

[2] R.K. DeLong, U. Akhtar, M. Sallee, B. Parker, S. Barber, J. Zhang, M. Craig, R. Garrad, A.J. Hickey, E. Engstrom, Characterization and performance of nucleic acid nanoparticles combined with protamine and gold, Biomaterials 30 (32) (2009) 6451–6459. Epub 2009 Sep 1.

[3] L.R. Brewer, M. Corzett, R. Balhorn, Protamine-induced condensation and decondensation of the same DNA molecule, Science 286 (5437) (1999) 120–123.

Additional Materials

[1] Aldevron, http://www.aldevron.com/products/dnas/gwiz/ (accessed 30.09.14)

[2] Rodney Boyer, Biochemistry Laboratory—Modern Theory and Techniques, 2nd ed., Pearson Prentice Hall, 2012, 2006, Chapter 6, Characterization of Proteins and Nucleic Acids by Electrophoresis.

[3] O. Gaal, G.A. Medgyesi, L. Vereczkey, Electrophoresis in the Separation of Biological Macromolecules, John Wiley and Sons, 1980.

[4] J. Sambrook, E.F. Fritsch, T. Maniatis, Molecular Cloning: A Laboratory Manual. Cold Spring Harbor Laboratory, 1982.

[5] Frederick M. Ausubel, Current Protocols in Molecular Biology, vol. 1, John Wiley and Sons, 1993, Chapter 2, Preparation and Analysis of DNA.

[6] G.D. Fasman, Handbook of Biochemistry and Molecular Biology, Proteins I, 3rd ed., CRC Press, 1976, pp. 183–203.

Notes

_____

_____

_____

_____

_____

_____

_____

_____

8

Qualitative Analysis of the Degradation of RNA via Ribonuclease A versus B

SAFETY AND HAZARDS

You should wear gloves while performing this lab. Whereas the RNA is harmless, the ethidium bromide (EtBr) stain, as used previously, is a mutagen/carcinogen, and all samples and materials exposed to or containing EtBr should be handled and disposed of properly. Electrophoresis also has hazards associated with it (see Lab 4 and Lab 7).

INTRODUCTION

It could be said that we have entered the "RNA era" of molecular biology. RNA function, along with RNA:protein interactions, has become yet another important piece of the puzzle in the molecular basis of life. With the discovery of pancreatic ribonuclease (RNase) by Jones in 1920, researchers have turned their attention toward the function and nature of ribonucleases [1]. Ribonucleases come in a variety of different classes. For instance, RNase H degrades RNA/DNA complexes [2], while RNase L is a 2′,5′-oligoisoadenylate synthetase-dependent ribonuclease [3]. The mammalian ribonuclease A superfamily is composed of RNase 1–13 [4]. RNase I, also known as RNase A, is the best studied ribonuclease to date. It was first isolated from bovine pancreas. RNase A was the first enzyme and third protein to have its amino acid sequenced correctly. This enzyme's stability has shown to be dependent on four disulfide bonds, while its catalytic potential is dependent on histidine 12 and 119. It has been proven that RNase A has a poly C > poly U > poly A specificity [5]. More recent studies have turned focus toward post-transcriptional modification.

RNase B, which has the same amino acid sequence of RNase A, contains a single glycosylation site at Asn 34, which has five glycoforms depending on the mannose content [6]. This experiment focuses on the time-course of RNA degradation and RNA analysis and compares the unglycosylated form of ribonuclease (A) to the glycosylated form (B).

EXPERIMENT 8

Equipment and Materials

- Torula Yeast RNA (Sigma–Aldrich)
- DEPC-treated RNase-free ddH$_2$O
- Bovine pancreas RNase A and B (Sigma–Aldrich)
- 1X TAE buffer
- Low-melting gel-grade agarose
- Gel electrophoresis equipment and documentation station with UV lamp and imager
- Bromophenol blue loading buffer (40% sucrose)

Methods and Procedure

Handle all samples and equipment while wearing gloves. Make stock concentrations of 2 μg/μL and 5 μg/μL of Torula Yeast RNA in small 0.5 mL presterilized Eppendorf tubes. Also, make stock concentrations of 1 μg/μL and 2 μg/μL of bovine pancreas RNase A and B in sterile water immediately prior to use and store on ice. Prepare a 2% agarose gel as directed previously in Lab 7. Include 1 μL of 1 mg/mL EtBr in gel to stain the RNA and visualize by UV on the gel documentation station.

RNase Degradation versus Time

Use Table 8-1 to establish the layout of your gel. All lanes also contain 3 μL of bromophenol blue loading buffer (40% sucrose). A total of 15 μL was pipetted into each well. The RNA and RNase stock used for this gel was 5 μg/μL and 1 μg/μL, respectively. Run the gel at 100 volts for 45 minutes, or until the loading dye has migrated halfway through the gel.

RNase Degradation A versus B

Use Table 8-2 to establish the layout of your gel. All lanes also contain 3 μL of bromophenol blue dye loading buffer. The RNA and RNase stock used for this gel was 2 μg/μL. Lane 2 uses RNase concentration of 1 μg/μL. The reaction was allowed to take place for 10 minutes before being pipetted into the wells. A total of 15 μL was pipetted into each well. Run the gel at 100 volts or for 20 minutes, or until loading dye is halfway between the lanes.

TABLE 8-1   Degradation Kinetics

Lane	1	2	3	4	5	6	7	8
RNase A (μL)	0	1	1	1	0	0	0	1
RNase B (μL)	0	0	0	0	1	1	1	1
RNA (μL)	2	2	2	2	2	2	2	2
H₂O (μL)	12	12	12	12	12	12	12	12
Time (min)	10	1	5	10	1	5	10	1

TABLE 8-2   Degradation with Increasing Amount of RNase A

Lane	1	2	3	4	5	6	7	8
RNA (μL)	2	2	2	2	2	2	2	2
RNase A (μL)	0	1	1	2.5	5	7.5	10	12.5
H₂O (μL)	13	12	12	11	8	6	3	1
Lane	1	2	3	4	5	6	7	8
RNA (μL)	2	2	2	2	2	2	2	2
RNase B (μL)	0	1	1	2.5	5	7.5	10	12.5
H₂O (μL)	13	12	12	11	8	6	3	1

RESULTS AND DISCUSSION

A typical set of results is shown below.

Figure 8-1 depicts the degradation of RNA over time at 1, 5, and 10 minutes of incubation. Qualitatively, it appears that RNase A as seen in lanes 2–4 shows a fainter band, representing a greater magnitude of RNA catalysis in comparison to RNase B (lanes 4–7). In lane 8, which contains equal amounts of A:B, there appears to be greater catalysis than in lanes 5–7 but not as significant as lanes 3 and 4.

Figure 8-2 shows the degradation of RNA with increasing concentrations of RNase. This image shows almost complete catalysis of RNA with increasing concentrations of both enzymes. The top band represents RNase A with almost complete catalysis of the RNA when the ratio of RNase to RNA is 6:1. The bottom band represents RNase B with similar but somewhat miniscule differences. In lane 3, there is significant difference in RNase A versus lane 2. However, for RNase B, there is almost no significant change in the two lanes.

Based on your results, summarize them for your conclusion section and consider the following questions from the problem set in writing up your notebook for this experiment.

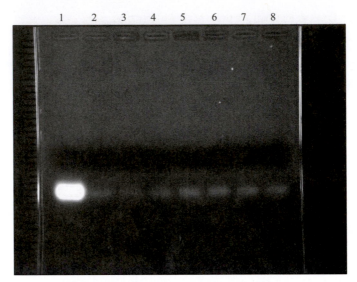

FIGURE 8-1   Kinetics of RNA degradation

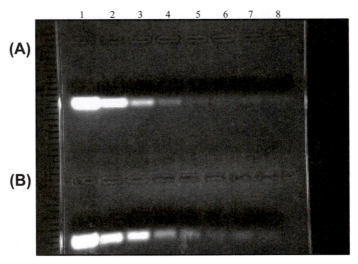

FIGURE 8-2   Dependence of RNA to RNase ratio

PROBLEM SET

1. Which enzyme is more active?
2. Was the RNA degraded more over time with RNase A or B?
3. Since the two enzymes differ only in having a/an (un)glycosylated amino acid residue, what can be inferred about how the glycosylation affects enzymatic activity?

4. From a biomolecular interactions perspective, what are some plausible explanations for this?

5. How would you test these hypotheses experimentally?

References

[1] Krystal Worthington, A. Ribonuclease, Worthington Enzyme Manual, Worthington Biochemical Corporation, June 23, 2014.

[2] Rosta, Edina. Two-metal ion catalysis by ribonuclease H. Department of Chemistry, King's College, London, n.d.

[3] A. Zhou, Interferon action and apoptosis are defective in mice devoid of 2',5'-oligoadenylate-dependent RNase L, EMBO J. 16 (21) (1997) 6355–6363.

[4] Soochin Cho, J. Beintema Jaap, Jianzhi Zhang, The ribonuclease A superfamily of mammals and birds: identifying new members and tracing evolutionary histories, Genomics 2005 85 (2) (June 23, 2014) 208–220.

[5] Ronald T. Raines, Ribonuclease A, Chem. Rev. 98 (3) (1998) 1045–1066.

[6] Heidi C. Joao, Raymond A. Dwek, Effects of glycosylation on protein structure and dynamics in ribonuclease B and some of its individual glycoforms, Euro. J. Biochem. 218 (1) (1993) 239–244.

Notes

Preparation of a Fluorescently Labeled Liposome and Its Analysis by Fluorescence Microscopy

SAFETY AND HAZARDS

The solvents used in this experiment, such as chloroform ($CHCl_3$), are toxic and should be handled with great care in the chemical hood, and care must be taken to avoid skin contact, inhalation, or ingestion. As always, you should check the MSDS sheets in advance of the lab and become familiar with any safety and health issues before conducting the experiment. If a rotary evaporator is used (preferred), your instructor will instruct the class in its safe operation and supervise its use. The sonicator may generate high-pitched noise, which may cause damage to your hearing. Wear earplugs when you work with the sonicator.

INTRODUCTION

The purpose of this lab is to prepare a liposome as a model membrane structure. In doing so, you will also gain an initial appreciation for the processes called good manufacturing practice, or GMP, typically used in many biotechnology companies. GMP is involved in producing biologically based products such as liposomes for use in humans.

Liposomes have a variety of applications in research and development and are widely used in drug and gene delivery [1], where they can be used to incorporate substances in the interior or associated with the membrane for association with or entry into cells. For example, some very potent anticancer drugs are poorly water soluble; thus, liposomal formulations can greatly improve the drugs' safety and efficacy. The classic well-known example of this is the drug doxorubicin and its liposomal formulation known as Doxil.

A company producing materials such as liposomes destined for use in humans would need to comply and strictly adhere to the U.S. Food and Drug Administration's (FDA) guidance and regulations. The liposomes would be produced under what is called good manufacturing procedures. Liposome manufacture would occur in a clean room setting, and the lipids used to make the liposomes, any solvents, and any other material used in the process would be made with only the purest ingredients. These components and any glassware, pipettes, and any material that came in contact during the process would need to be extremely clean, free of any contaminants, and sterile. The term *sterile* here refers to being free from any microbial contamination such as yeast and bacteria, which, in addition to the organism, might otherwise introduce or contain other dangerous agents such as lipopolysaccharide or endotoxin. In today's lab, you model GMP somewhat, so that you can gain an appreciation for the rigor and control that scientists and research and development technicians who are employed in the industry must use in these areas.

There has been much research in the past 20 or so years on liposomes, and there is a vast literature on them. On one level, they can be categorized into anionic, cationic, or neutral-surface-charge type liposomes. Typically, liposomes are formulated with a charged or polar lipid type, such as phospholipids. For stability, they are combined with a more neutral hydrophobic lipid (cholesterol is a common choice). On another level, an exciting and more modern approach is to attach or include in the liposome some type of targeting element, which may give liposomes some ability to selectively home for a particular tissue or organ (i.e., brain) or tumor in the case of cancer treatment. Thus, peptide, receptor, and other biomolecule inclusion in the liposome—ideally its exterior—is actually possible. In this way, liposomes can be recognized by a specific cell type and taken in selectively. In this case, it is advantageous to know something about the organ or cell type you are trying to target and whether a specific ligand, protein, or type of lipid in the liposome might help accomplish this [2].

When liposomes are combined with these proteins and other membrane-associating biomolecules, you may begin to think about them in terms of potential model membrane structures. At its most fundamental nature, the plasma membrane is a lipid bilayer. Other molecules, proteins (integral or peripheral), carbohydrates attached to proteins (glycolproteins), and lipid-protein conjugates (lipoproteins) make up a minority of the rest of the membrane. Thus, many researchers have begun to use liposomes as model cell membrane structures with which to study the structure of cell surfaces and interactions at the membrane; this can drive important biological functions such as vesicle fusion, secretion, receptor activity, and others. The important interactions between liposomes and cellular structures may involve, in addition to drug delivery, cell toxicity (cytotoxicity), viability, signaling transduction and many others.

At first glance, the preparation of a model liposome membrane seems relatively straightforward. Simply dissolve the lipid in an organic solvent, create a thin film of it around a glass flask, and then introduce water and some sort of physico-mechanical force (sonication) to force the lipid into the context of the water bulk. The liposomes will spontaneously form, due primarily to hydrophobic forces, exclusion of water, and minimizing exposure of the polar water molecules with the nonpolar parts of the lipid. While, to a certain extent, this is true, suffice it to say that the preparation of uniform-size homogenous liposomes has been a great challenge up until quite recently [3].

Using a micro-rotary evaporator technique and by substituting a neutral triacylglyceride (TAG) molecule for the phospholipid, along with inclusion of cholesterol at a 1:1 (wt:wt) ratio, our research group has recently succeeded in producing neutral surface liposomes [4]. These are called *niosomes*. The lipids are dissolved in a minimum amount of an organic solvent (chloroform) and co-evaporated to a dry, thin film. The lipids are sonicated in the presence of sterile phosphate buffered saline (PBS) with the spontaneous formation of liposomes. These generally range from one to several microns and can be viewed by light microscopy, as shown in Figure 9-1.

As shown in the right panel of the figure, a small amount of a fluorescently tagged lipid, such as cholesterol-conjugated fluorophore, can be incorporated into the liposome, allowing their visualization by fluorescence microscopy.

Fluorescence is a property of molecules whereby, similarly to absorbance, when they absorb light at a particular wavelength (excitation), they give off light at a slightly longer wavelength (emission). Fluorescence microscopy is one of the most common techniques used in molecular biology and biochemistry today. Typically, cells are labeled with one or more fluorescent labels, which could be proteins—for example, green fluorescent protein (GFP) or red

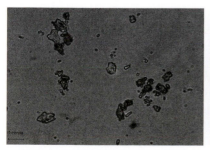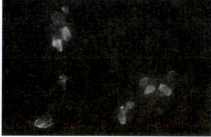

FIGURE 9-1 Sample image of noisome by light microscopy (left panel) or fluorescence microscopy (right panel)

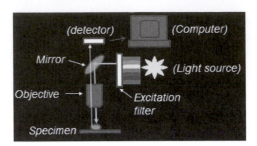

FIGURE 9-2   Basic setup of a fluorescent microscope

fluorescent protein (RFP)—or other fluorophores. The microscope has the ability to produce light in a specific wavelength range that a particular type of fluorescent molecule (fluorophore) selectively absorbs and becomes excited by; it then gives off or emits light most commonly in the green range (e.g., fluorescein) or red range (e.g., Rhodamine or Texas Red), which can be viewed or detected by a CCD (charged coupled device) camera. A computer with built-in software is then able to build an image and project it. A schematic of the basic working parts of a fluorescent microscope is shown in Figure 9-2.

In summary, in this experiment you produce a liposome containing a small amount of fluorescent cholesterol so that you are able to observe it by using fluorescence microscopy, as shown in Figure 9-1 and Figure 9-2. You conduct the experiment under mock GMP conditions, attaining signoff from your instructor at each step of the process, and assess the sterility of your product in a typical microbial growth agar.

EXPERIMENT 9

Equipment and Materials

- Tripalmitin
- Cholesterol
- Chloroform ($CHCl_3$)
- Clean round-bottom (RB) flask
- Sterile PBS buffer
- Fluorescent microscope, slides, and cover slips
- Sonicator
- Fluorescent cholesterol—NBD Cholesterol (Invitrogen/Molecular Probes N1148) or Bodipy cholesterol (Molecular Probes 5421563C11)
- Rotary evaporator with vacuum line (optional)
- Sterile agar plates for microbial growth and incubator
- Analytical balance (calibrated and certified prior to use)
- General lab equipment (pipettemen, tips, Eppendorf microfuge tubes, small weigh boats and paper, clean spatula, and glassware)

Methods and Procedures

This lab is to be conducted under mock GMP conditions in which you and your partners should initialize the lines provided below after completing each individual step. Also, you should obtain the instructor's signature prior to and after completing the simulated quality control analysis of the liposome.

1. Sign out from your instructor small amounts of each lipid type, tripalmitin and cholesterol as dry powders.
   _____ (Initial here when complete.)
2. Weigh out 5–10 mg of each lipid on the analytical balance on a clean weigh paper or small boat; then transfer and combine them in the round-bottom flask.
   _____ (Initial here when complete.)
3. Working in a clean chemical hood and using a clean glass pipette, transfer a small volume (1 to 2 mL) of chloroform to the RB flask; then gently swirl the solvent around the sides of the flask to dissolve the lipids. Add 5–10 μL of the fluorescent cholesterol (0.5 mg/mL solution in $CHCl_3$ or methanol).
   _____ (Initial here when complete.)
4. Connect to a rotary evaporator with a vacuum line, and while the flask is rotating, evaporate the solvent under the vacuum (this process should take only 5–10 minutes). Alternatively, swirl the flask by hand with a gentle stream of filtered air or nitrogen, the goal being to leave a thin film of lipid on the side of the RB flask.
   _____ (Initial here when complete.)
5. Once the thin film is apparent but before the lipids crystalize out of the solution, quickly transfer approximately 10 mL sterile PBS. Cap or parafilm-seal the top in a sterile hood, vortex the RB vigorously for 1–2 minutes, and then place into a sonicator bath and sonicate for 5–10 minutes.
   _____ (Initial here when complete.)
6. After sonication, vortex the suspension vigorously for about 30 seconds. Then using a 10–20 μL pipetteman, mix up and down rapidly, transfer onto a slide, and mount the cover slip.
   Manufacturing phase complete _____ (Instructor signature)

Quality Control Analysis

In this phase of the experiment, the goal is for your group to visualize the liposome first by light microscopy followed by fluorescence microscopy and then to streak a plate to be sure that the liposome is sterile.

1. Microscopy: With help from your instructor, place the slide onto the specimen stage of the microscope, as shown in overview in Figure 9-2. Bring the sample into focus first with bright light using a low-power objective, and once a group of liposomes is observed, switch to a high-power objective (40X or 100X). Have your instructor check to be sure that you are visualizing liposomes rather than air bubbles by switching to fluorescence mode. The image should appear similar to Figure 9-1.
2. Sterility test: In a sterile hood and using a flame-sterilized loop or another presterilized spatula or pipette tip, streak or transfer 20 to 100 μL of the liposome onto a sterile agar plate. Place the agar plate in the incubator and observe 24 hours or up to a week later to be sure the liposome is free of microbial contamination.
   Quality control/analysis phase complete _____
   (Instructor signature)

RESULTS AND DISCUSSION

1. Record your observations, make sketches, and attach any photographs or images from your data in your notebook.
2. Consider developing a summary table to organize your observations and results.

PROBLEM SET

In writing up your discussion, you might reflect on some of the following questions and incorporate them into your discussion:

1. What are the Chemical Abstract Service (CAS) numbers for the lipids used today?
   _____ (Tripalmitin)
   _____ (Cholesterol)
2. What is the mole-to-mole ratio of the lipids used, including the fluorescently labeled cholesterol?
3. Design an experiment for which the goal or end result is to confirm or test the mole-to-mole ratio of lipids within a given liposome test batch.
4. How you would expect the temperature or other conditions/ parameters during the liposome preparation stage to affect the outcome of your experiment?
5. In your research, suppose you are interested in using a putative liposome to deliver a drug to a particular tissue (e.g., liver or lung). What additional or different components would the liposome possess, and how would you incorporate them or associate them with the liposome?

References

[1] J. Smith, Y. Zhang, R. Niven, Toward the development of a non-viral gene therapeutic, Adv. Drug Deliv. Rev. 26 (2–3) (1997) 135–150.

[2] S. Rai, R. Paliwal, B. Vaidya, K. Khatri, A.K. Goyal, P.N. Gupta, S.P. Vyas, Targeted delivery of doxorubicin via estrone-appended liposomes, J. Drug Target 16 (6) (2008) 455–463.

[3] Renee R. Hood, Don L. DeVoe, Javier Atencia, Wyatt N. Vreeland, Donna M. Omiatek, A facile route to the synthesis of monodisperse nanoscale liposomes using 3D microfluidic hydrodynamic focusing in a concentric capillary array, Lab Chip 14 (2014) 2403–2409.

[4] R.K. Delong, A. Risor, M. Kanomata, A. Laymon, B. Jones, S.D. Zimmerman, J. Williams, C. Witkowski, M. Warner, M. Ruff, R. Garrad, J.K. Fallon, A.J. Hickey, R. Sedaghat-Herati, Characterization of biomolecular nanoconjugates by high-throughput delivery and spectroscopic difference, Nanomedicine (Lond.) 7 (12) (2012) 1851–1862.

Notes

Studying Cell-like Structures with Liposome, DNA, and Protein

SAFETY AND HAZARDS

The solvents used in this experiment, such as chloroform ($CHCl_3$), are toxic and should be handled with great care in the chemical hood, and care must be taken to avoid skin contact, inhalation, or ingestion. As always, you should check the MSDS sheets in advance of the lab and become familiar with any safety and health issues before conducting the experiment. If a rotary evaporator (or rotovap) is used (preferred), your instructor will instruct the class in its safe operation and supervise its use. The sonicator may generate high-pitch noise, which may cause damage to your hearing. Wear earplugs when you work with the sonicator.

INTRODUCTION

All organisms have an organization at the cellular level. As you have learned in this book, at the subcellular level, every cell is made up of the four main classes of biomolecules—lipids, proteins, nucleic acids, and carbohydrates—working synergistically together to regulate the molecular basis for life. Because some forms of RNA are well known to be catalytic, and "DNAzymes" have also been artificially synthesized [1], many scientists believe the primitive, early life forms may have been some simple nucleic acid structures bound or enclosed by a membrane to protect it. In any case, as the ultimate example of biomolecular structure, in this lab the aim is to create such an entity that is a membrane-enclosed nucleic acid and to stain or label each structure and confirm it by using fluorescence microscopy. As an additional exercise if time permits, you can use a simple, safe, and well-described protein such as albumin or protamine to interact with the membrane and stain it colorimetrically for observation. Such lipid-protamine-DNA (LPD) structures were first described by Leaf Huang's research group at the University of North Carolina and have exciting possibilities for the delivery of DNA and for gene therapy [2–3]. Dr. DeLong's research group has made fluorescently tagged liposome containing phospholipid and studied their interaction with proteins via high-throughput chromophoric or fluorescence-difference spectroscopy [4]. Therefore, you might like to explore these types of applications as an additional experiment or as it relates to your own research, such as incorporation of proteins into the interior or into the membrane itself, etc.

EXPERIMENT 10

Equipment and Materials

- Cholesterol
- Dioleoylphosphatidylcholine (DOPC) [Sigma–Aldrich or Avanti polar lipids]
- Chloroform ($CHCl_3$)
- Clean round-bottom (RB) flask
- Sterile PBS buffer
- Fluorescent microscope, slides, and cover slips
- Sonicator
- Fluorescent cholesterol—Bodipy cholesterol (Molecular Probes 5421563C11)
- Rotary evaporator with vacuum line (optional)
- DAPI stain
- Analytical balance (calibrated and certified prior to use)
- General lab equipment (pipettemen, tips, Eppendorf microfuge tubes, small weigh boats and paper, clean spatula, and glassware)

METHODS AND PROCEDURES

Liposome Protocol

1. Turn on the rotary evaporator and allow it to warm up for about 15 minutes. Set the temperature to 35°C–40°C. Also turn on the water hose to prevent the machine from overheating.
2. Measure about 4 mg of cholesterol and place in a round-bottom flask.

3. Measure about 4–5 mg of phosphatidylcholine and add it to the round-bottom flask. Keep the PC on ice when not measuring. (The substance will become oily as it begins to heat up at room temperature, so you might need to wash it off the spatula using the chloroform.)

4. Add 2–3 mL of chloroform to the round-bottom flask to dissolve the lipids.

5. Screw the flask onto the rotary evaporator; then turn on the vacuum and the rotation knob. Alternatively, swirl by hand in a gentle stream of air or nitrogen gas to evaporate lipids to thin film.

6. If you are using the rotovap, watch the flask for several minutes. Look for evaporation of all the chloroform and for a thin film to form around the inside of the round-bottom flask.

7. Add 10 mL of 1X PBS and 600 µL of PCR DNA, lambda phage DNA, or plasmid DNA (0.5–1 mg) and vortex the sample vigorously for 30 seconds to 1 minute.

8. Set up the sonicator with a beaker stand to hold the flask in the sonicator. Allow this to sonicate for 20–30 minutes.

9. After 30 minutes, aliquot all the contents of the flask into 1.5 mL centrifuge tubes. Centrifuge the tubes for 5 minutes at 10,000 g (if liposomes have formed, you should see a precipitate in the bottom of the centrifuge tube after each centrifugation).

10. Remove the supernatant and discard into a beaker. Wash the precipitate with 1X PBS and centrifuge. Repeat this step five times. (Doing so helps remove any DNA that has failed to be enveloped in the liposomes.)

11. After the final wash, resuspend the liposomes in the tube using the vortex. Transfer all the resuspended liposomes into one cell culture tube and vortex again.

DAPI Staining

1. To stain the DNA within the liposomes, use DAPI. Do two separate stains on the same slide.

2. On the right side of the slide, transfer 10 µL of liposome and 5 µL of DAPI (3 µM) stain. Once both are on the slide, mix by sucking up and down in the pipette several times.

3. On the left side of the slide, transfer 5 µL of liposome and 10 µL of DAPI. Mix the liposome and DAPI with the pipette (doing so keeps the total volume of the two stains the same).

4. Cover both sides with cover slips.

5. Under the guidance of your instructor, view the slide under the fluorescent microscope to determine if DNA has been enclosed within the liposomes.

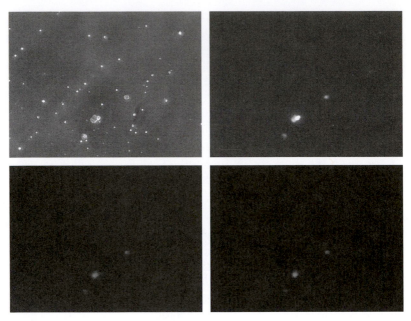

FIGURE 10-1 Sample images of liposome by light and fluorescent microscope. Top left: bright field image of nonhomogenous liposomes; top right: DAPI stain of DNA entrapped by liposomes; bottom left: green fluorescent signal of a fluorescent-tagged liposome (NBD-cholesterol) incorporated in the liposome; bottom right: red fluorescent signal of another fluorescent-tagged liposome (Bodipy-cholesterol) incorporated in the liposome. Scale bar: 10 μM

Figure 10-1 shows a sample picture of liposomes taken by a fluorescent microscope.

RESULTS AND DISCUSSION

1. Record your observations, make sketches, and attach any photographs or images from your data in your notebook.
2. Consider developing a summary table to organize your observations and results.

ADDITIONAL EXPERIMENT

You may consider binding albumin or protamine to the liposome and staining the lipid-protamine-DNA surface with a dye such as bromophenol blue or trypan red, in which case the protein-associated liposome can be viewed by using light microscopy or evidence of the structure obtained by shifts in the colorimetric spectrum of the dye [4].

References

[1] C. Wilson, J.W. Szostak, Isolation of a fluorophore-specific DNA aptamer with weak redox activity, Chem. Biol. 5 (11) (1998 Nov) 609–617.

[2] J. Dileo, R. Banerjee, M. Whitmore, J.V. Nayak, L.D. Falo Jr., L. Huang, Lipid-protamine-DNA-mediated antigen delivery to antigen-presenting cells results in enhanced anti-tumor immune responses, Mol. Ther. 7 (5 Pt 1) (2003 May) 640–648.

[3] S. Li, M.A. Rizzo, S. Bhattacharya, L. Huang, Characterization of cationic lipid-protamine-DNA (LPD) complexes for intravenous gene delivery, Gene. Ther. 5 (7) (1998 Jul) 930–937.

[4] R.K. Delong, A. Risor, M. Kanomata, A. Laymon, B. Jones, S.D. Zimmerman, J. Williams, C. Witkowski, M. Warner, M. Ruff, R. Garrad, J.K. Fallon, A.J. Hickey, R. Sedaghat-Herati, Characterization of biomolecular nanoconjugates by high-throughput delivery and spectroscopic difference, Nanomedicine (Lond.) 7 (12) (2012 Dec) 1851–1862.

Notes

_____

_____

_____

_____

_____

_____

_____

_____

_____

_____

_____

_____

_____

_____

_____

_____

_____

_____

Index

Note: Page numbers followed by "f" and "t" denote figures and tables, respectively.

A

Acetonitrile (CH₃CN), 21
Agarose gel, 69, 69f
   for electrophoresis, 69, 70f
Amine (NH₃⁺), 22
Amino acids
   discussion, 31–32
   equipment highlight, 26–27
   methods and procedure, 28–31
   problem set, 32–33
   results and observations, 31
   safety and hazards, 21
   side-chain functional group (R)
       categories, 22, 23t–24t
   solubilizing, 28, 29t
   structure, 21, 22f
   technical review, 25–26
Annealing temperature, 62

B

Bead technology, 38–39
Beer-Lambert law, 52–53
Bench top centrifuge, 37, 38f
Biomolecular laboratory
   equipment highlight, 6–8
   experiment, 8
   methods and procedure, 8–9
   problem set, 10
   results and discussion, 9
   safety and hazards, 1
   technical review, 2–4
Buffers
   equipment highlight, 16–17
   experiment, 17
   and pH, 15
   problem set, 18
   results and discussion, 18
   safety and hazards, 13
   technical review, 14–15

C

Carbohydrates, 45
Carboxylic acid (COO⁻) groups, 22
Cell lysis, 37

Centrifugation, 37
Chloroform (CHCl₃), 87
Chromatography, 25
CI. *See* Confidence intervals (CI)
Column chromatography, 38–39
Confidence intervals (CI), 4
Cysteine, molecular structure of, 25, 25f

D

DAPI staining, 89–90
Data analysis software, 3
Data-handling process, 50, 51t
Deoxyribonucleic acid (DNA)
   experiment, 88
   liposomes, 90, 90f
   methods and procedures, 88–90
   polymerase, 60
   results and discussion, 90
   safety and hazards, 87
   template, 60
Dilutions, 14–15, 49–50
Double-distilled, deionized water
   (ddH₂O), 14

E

Electrophoretic mobility shift assay
   (EMSA)
   equipment highlights, 69
   experiment, 69–71
   problem set, 71–72
   results and discussion, 71
   safety and hazards, 67
Ethidium bromide (EtBr), 68, 73

F

Fluorescence, 81–82

G

Gel electrophoresis, 39
   equipment, 69
Genes encoding, 35
Glucose measurement
   equipment highlight, 54

experiment, 54–56
problem set, 56
results and discussion, 56
safety and hazards, 45
technical review, 49–53
Glucose-6-phosphate dehydrogenase
    (G6PDH), 49
Green fluorescent protein (GFP), 68

H
Hexokinase, 49
Hydrophobic interaction chromatography
    (HIC), 26

I
Ion exchange (IEX), 38–39

L
Liposome, 90, 90f
    experiment, 82–84
    problem set, 84–85
    protocol, 88–89
    results and discussion, 84
    safety and hazards, 79
Liquid transferring tools, 7–8, 7t

M
Mass balance, 6
    instrument, 6, 6f
Mean, 3
Methanol (CH$_3$OH), 21
Micropipettes, 7–8
Microplate reader, 54
Micro-rotary evaporator technique, 81
Microscopy, 84
Molarity, 14
Molar absorption coefficient, 52–53

N
Nucleotides, 60

P
pH meter, 16, 16f
Phosphate buffered saline (PBS), 13–14, 81
Polyacrylamide gel electrophoresis
    (PAGE), 13–14, 39
Polymerase chain reaction (PCR)
    buffers, 60–61, 61t
    cycles, 61, 61t
    experiment, 64–65
    mixture preparation, 64, 65t

problem set, 66
results and discussion, 65
safety and hazards, 59
technical review, 62–63
Precipitation, 37–38
Primers, 60
    design, 62
    length, 62
Proline, molecular structure of, 22, 25f
Protamines, 68, 68f

Q
Quantitative liquid transfer (pipetting), 7

R
Range, 4
Rapid purification
    safety and hazards, 35
    technical review and equipment
        highlights, 36–39
Ribonucleic acid (RNA)
    degradation of, 74–75, 75t, 76f
    experiment, 74
    with increasing concentrations of RNase,
        75, 76f
    problem set, 76–77
    results and discussion, 75
    safety and hazards, 73

S
Single-stranded binding (SSB) protein, 67
Size exclusion chromatography (SEC),
    38–39
Sodium dodecyl sulfate (SDS), 13–14
Standard curves, 50
Standard deviation (SD), 4
Standard pipettes, 7
Statistical analyses, 3
Sterility test, 84
Student's t-test, 4, 5t

T
Thin layer chromatography (TLC), 26–27,
    27f
    eluting, 30–31
    spotting, 30
Triacylglyceride (TAG), 81
Trichloroacetic acid (TCA), 37–38

U
UV/Vis spectrophotometry, 50–52